Dirty Virtues

Dirty Virtues

The Emergence of Ecological Virtue Ethics

Louke van Wensveen

Published 2000 by Humanity Books, an imprint of Prometheus Books

Inquiries should be addressed to
Humanity Books
4501 Forbes Boulevard, Suite 200,
Lanham, Maryland 20706
www.rowman.com
Distributed by NATIONAL BOOK NETWORK
800-462-6420

Library of Congress Cataloging-in-Publication Data

Wensveen, Louke van.
Dirty virtues : the emergence of ecological virtue ethics / Louke van Wensveen.
p. cm.
Includes bibliographical references and index.
ISBN 978-1-57392-649-2 (cloth : alk. paper)
1. Environmental ethics. 2. Human ecology—Religious aspects. 3. Human ecology—Philosophy. I. Title.
GE42.W455 1999
179'.1—dc21 99-045065
CIP

for my children, Derek and Ursula
who made me look to the future

CONTENTS

PREFACE

LIKE MANY BOOKS, THIS is a book with a personal genesis. My parents have told me that I was always a "good girl." So good that it worried them. Given this background, my professional interest in virtue ethics—the systematic reflection on goodness and character—comes hardly as a surprise, I suppose. Yet it has taken me in a surprising direction. I will not comment on whether practicing the subtleties of academic reflection have turned me into a "bad girl," but it has certainly led me on to the track of an uncommon batch of virtues and vices.

It all started when I became pregnant and faced the prospect of raising children in Los Angeles. Contemplating my city's dubious environmental qualities, I experienced a combination of panic, denial, anger, and a "can do" optimism. In the end I settled for making some lifestyle changes and turning the direction of my career. As an ethicist I figured the least I could do is focus my research on ecological questions.

So while I was waiting for my firstborn to arrive, I dove into the burgeoning world of ecoliterature and discovered something I did not expect: lots of virtues. Some were familiar, others quite different from what I had ever heard about or seen in others or tried to cultivate myself. This book is the result of my fascination with the rich and effervescent virtue language that is turning up in the literature of environmental philosophy and theology, a language rooted in the practical commitments of ecologically minded people.

The tenor of the book is hermeneutical. Part One offers an overview of ecological virtue language from three different but complementary interpretive venues. I begin by exploring general characteristics of the language (chapter 1). Next I place ecological virtues and vices in the context of the Aristotelian tradition, highlighting similarities as well as differences (chapter 2). Then follow two chapters in which I focus on the details and variations of ecological virtue language by means of in-depth interpretations of the work of two leading authors, Murray Bookchin and Thomas Berry (chapters 3 and 4).

Interpretation is, of course, to a degree a constructive effort, and in Part Two the constructive dimension becomes my focus. Here I propose a number of criteria intended to clarify what makes for genuine virtue and true vice in an ecological age. In five chapters I variously draw on Aristotelian and Christian traditions, as well as on feminist and psychological discourse, in order to gain a critical perspective aimed at sharpening the concepts and conventions of ecological virtue language. My hope is thus to contribute to the construction of virtue theories that are appropriate for an ecological age as well as connected to some of the better aspects of Western thought.

I would like to thank Linda Zagzebski and Bernard Reginster for their sustained interest in the project and for the support of their friendship. My colleagues Thomas Rausch, S.J., Herbert Ryan, S.J., John Popiden, Jeffrey Siker, and James Hanink all read drafts of chapter 6 and provided many helpful comments. Christopher Chapple graciously invited me to write a concluding essay for the collection *Ecological Prospects: Scientific, Religious, and Aesthetic Perspectives*, parts of which have been reworked into chapters 1 and 2. I am also grateful to the members of the Society of Christian Ethics who have given me feedback on presentations related to chapters 1, 6, and 8.

My excellent student research assistant, Kim Christensen, has done much of the tedious work involved in culling ecological virtues and vices from a wide range of literature. Her precision, independent judgment, and genuine interest have been a great source of support for me.

Two institutions have facilitated the writing of this book. I would like to thank CEKUN, the ethics research center at the University of Nijmegen, the Netherlands, for commissioning a paper that has been incorporated into chapter 9 and for supporting the final stage of writing the manuscript. I am also very grateful to Loyola Marymount University for awarding me three Summer Research Grants, funding for research assistance, as well as the 1994–95 Casassa Chair of Social Values.

Scholarly lectures can change lives. In February of 1989, when I was expecting my first child, Larry Rasmussen addressed the Pacific section of the Society of Christian Ethics with a strong appeal for ecological awareness. He is the intellectual catalyst behind this book.

I am most indebted to those who are closest to me. My father and stepmother have provided me with an intellectual's dream: friendship and great food during many months of writing. My mother welcomed me home summer after summer, only to have to relinquish her daughter to a computer and a pile of books. Yet she kept me honest by pulling me away for long walks in the new national park near her home, teaching me the practical knowledge of local ecology that has been handed down for generations. Back in Los Angeles, I have tried to cultivate the same attentiveness on my walks to and from work, so that even the exotic suburban trees and shrubs and flowers I pass every day have now become close to me. I must acknowledge that they have supported me over the years. Still, for an uprooted foreigner no support could have been as powerful as the support of another uprooted foreigner, my partner Eduard Hovy, who has tended the spirit of wonder and hope that has enabled me to finish this book.

Part 1 One

ECOLOGICAL VIRTUE LANGUAGE

Chapter 1 One

THE EMERGENCE OF ECOLOGICAL VIRTUE LANGUAGE

THE WORD "VIRTUE" HAS an old-fashioned ring to it. The word "vice" perhaps even more so. When as Westerners we hear these terms, we may think of professors belaboring the significance of the ancient Four Cardinal Virtues, or priests sermonizing on the perennial Three Theological Virtues and the Seven Deadly Sins. Even if we cannot exactly remember what they all are, we know these virtues and vices represent tradition, and so they inspire us with awe, or rather with resentment, or perhaps with an uneasy mixture of both. Then, of course, we may also think we have moved entirely beyond virtue and vice—either in skeptical rejection of tradition, or for lack of familiarity with it. Even so, "virtue" and "vice" will sound old-fashioned and hardly relevant to a book on the future-oriented and somewhat countercultural topic of ecological ethics.

I will not retort with an apology on the surprising relevance of established Western virtue traditions in an ecological age. Yet I will draw attention to a particular type of moral language that pervades the writings of those who seek to respond to the environmental crisis. In this mushrooming ecological literature, we are encouraged to care for our bioregions, to respect trees, to show compassion for the suffering of animals, to be humble and wise in the use of technology, to be frugal and creative in the use of limited resources, and to have hope in the face of impending global disaster. Conversely, we are warned to avoid the arrogance of anthropocentrism, to stop being cruel in our treatment of animals, to admit that we habitually project our fears onto nature, and to put a halt on our greed and the resulting manipulative exploitation of natural resources.

What would be an appropriate name for this language? The term "virtue language" (which includes vices as well) does seem appropriate. But then somehow we must indicate that we are not simply dealing with a rehash of tradition. Perhaps "ecological virtue language" is most descriptive, if somewhat mundane. I will use this expression in the remainder of the book.

Personally I like to think of the virtue language that ecologically minded people tend to use as "dirty virtues"—"dirty" because this language expresses a preoccupation with the earth (read: dirt = soil), and also because many ecological virtues would have been considered not particularly praiseworthy, or even vicious, during most of Western history (read: dirty = bad, taboo).

What does this dirty virtue language look like? The examples I have given immediately speak to the imagination. Respect for nature, caring, frugality, and hope clearly must be important attitudes in an ecological age. And arrogance, cruelty, and greed clearly must be avoided. Yet what exactly do we mean when we use these terms? Ecoliterature contains many thoughtful passages that relay the flavor of ecological virtues and vices. We also discover in this literature a much wider range of virtues and vices than we might expect. To give an idea of this richness in flavor and range, let me share some passages that describe ecological virtues:

> Re-specting nature literally involves "looking again." We cannot attend to the quality of relations that we engage in unless we know the details that surround our actions and relations.[1]

> The proper response to an emerging sense of the vastness and opacity of "wild connection" is not a (further) retreat into our own small realms of (supposed) transparency. Instead: fascination, interest, humility. Bowing before the mysteries of the world, entering a kind of wild etiquette.[2]

> Adherents to voluntary simplicity, those who content themselves with no more than they need, can now be found in significant numbers in most Western industrial societies and in some other societies as well. Deliberately abandoning the frenetic pursuit of material goods in favor of a simpler life-style, they are involved in recycling, home gardening, biking to work, and the rebellion against conspicuous consumption and planned obsolescence.[3]

> Awe and amazement are the results of a rich creation story, and the awe we feel should encompass our very selves, since every self is part of the unfolding creation story. We feel our interconnection with other creatures and peoples on this surprising planet in this amazing universe of one trillion galaxies, each with 200 billion stars.[4]

> I would define an ethical position most simply as this: acting to the best of one's ability from a sensibility that simultaneously knows and values oneself as an individual; is compassionate through identification with human and nonhuman others and caring about others' lives and well-being; and is creative, undergoing self-transformation through cultivating a relation to collectives ranging from human families to the planetary community.[5]

The richness of this language, and the fact that it flows from so many mouths, has inspired me to compile a catalogue of ecological virtues and vices—much as biologists record the variety of flora and fauna they encounter. This catalogue can be found in Appendix A, and I encourage my readers to dwell there for a while.

I have yet to come across a piece of ecologically sensitive philosophy, theology, or ethics that does not in some way incorporate virtue language. Ecological virtue language turns up in the writings of social ecologists as well as deep ecologists, bioregionalists as well as animal rights activists, creation theologians as well as environmental philosophers, mainstream theologians as well as radical ecofeminists. Within these different strands, however, ecological virtue language seems to turn up especially when authors assume a hortatory, personal, reflection-filled mode of writing. This most often happens in their concluding statements. Conversely, if a source contains much technical language, whether scientific, philosophical, theological, or metaethical, virtue language tends to turn up much less frequently. Still, even though remnants remain of the once widespread attempt to avoid normative discourse in the context of "objective" discussions, virtues and vices often seem to elude such a ban, so that even in more technical pieces authors can unabashedly use terms such as "respect" and "manipulation."

At this point we may well ask: If such a rich variety of ecological virtues and vices exists in such a wide range of literature, why is this moral language not better known? Why, in fact, have some commentators even lamented the low profile of virtues in environmental ethics?[6] This curious situation becomes understandable, I think, if we consider that those who have initiated key *methodological* discussions in the field of environmental ethics have generally not (or not primarily) been virtue ethicists.[7] Discussions in environmental ethics have largely focused on rights, values, duties, principles, and consequentialist arguments. This emphasis has created a blind spot for a segment of ecological moral language that, quietly living between the often heated debates about the rights of trees or the intrinsic value of rocks, actually flourishes quite profusely.

But, we might protest, choice of method is political, so the low profile of ecological virtue language cannot be seen as an innocent oversight! Val Plumwood, one of the few authors who has recently promoted a virtue-based ecological ethic, develops this argument based on a feminist analysis of the public/private distinction:

> Rights seem to have acquired an exaggerated importance in ethics as part of the prestige of the public sphere and the masculine, and the emphasis on separation and autonomy, on reason and abstraction. A more promising approach for an ethic of nature, and also one much more in line with the

> current directions in feminism, would be to remove rights from the centre of the moral stage and pay more attention to some other less universalistic moral concepts such as respect, sympathy, care, concern, compassion, gratitude, friendship and responsibility. . . . These more local moral concepts, because of their dualistic construal as feminine, and their consignment to the private sphere as subjective and emotional, have been treated as peripheral and given far less importance than they deserve.[8]

I find Plumwood's thesis quite plausible.

I also believe other factors have contributed to the methodological "virtue blindness" of environmental ethicists. For one thing, consider the context of litigation, an important focus of reform environmentalism. Here virtue language has much less bite than the language of rights. (I imagine that appealing to a chemical company's love of nature in a court of law would be as effective as appealing to an ex-spouse's love of his or her children in a child custody case.) No wonder then that in the midst of heated litigation involving toxic cleanup, pollution control, and the destiny of forests, virtues have not received much attention.

Another reason for the virtue blindness of environmental ethicists may be that recent interest in virtue ethics is about as old as the field of environmental ethics itself. Both represent nodes of great creativity in the field of ethics. We can imagine them as seeds that sprouted relatively independently and that only now, as plants, are beginning to cross-pollinate. Such a scenario is quite plausible given the specialized nature of much academic reflection and the limited ability of scholars to explore various new developments simultaneously. (This is the kind explanation. The more suspicious version states that the two seeds belong to different species, one conservative and the other progressive, which is the reason why they do not, nor should they be made to, interact much.)

Most of all, and here I return to my opening thought, I think the methodological virtue blindness of environmental ethicists is a result of the fact that the terms "virtue" and "vice" carry the stigma of sounding old-fashioned, preachy, and self-righteous. It is one thing to say, "We should respect nature!"; but it is another thing to reflect on this statement and happily conclude that one is uttering virtue language. Those of us who count themselves among the still largely countercultural activists who cultivate respect and love of nature in their daily lives—in other words, those who provide the practical base of environmental ethics—will more likely experience shock at this realization. (And vice versa, the theorists who have recently helped to revive interest in virtue ethics—i.e., those who would be made very happy by the realization that they too are, in fact, uttering virtue language—are generally not the ones who broadcast a personal interest in cultivating respect and love of nature.)

Instead, people who live ecologically responsible lives, and the academics who reflect on their lifestyles, tend to describe their ways of being and acting using the term "attitude." They typically coin phrases such as: "an attitude of humility,"[9] or "a new attitude of caring for creation."[10] Conversely, they reject our "aggressive attitude towards the natural world,"[11] and "anthropocentric and greedy attitudes."[12] Occasionally, the term "habit" is used. For example, Joanna Macy identifies the "habits of suppression" that keep us from being honest with our feelings and hence from experiencing compassion. Macy also revives the idea of "practice" (a traditional synonym for the cultivation of virtue), by which she means "fortifying the mind and schooling its attitudes."[13]

In all of these passages, the air is thick with a consciousness of life as an effort to grow from vice into virtue. Yet the terms "virtue" and "vice" themselves are conspicuously absent. I can only explain this widespread absence as the result of a certain stigma that must be attached to the *idea* of using virtue language. Listen again, for example, to Joanna Macy:

> Please note: virtue is not required for the emergence of the ecological self! This shift in identification is essential to our survival at this point in our history precisely *because* it can serve in lieu of "ethics" and "morality." Moralizing is ineffective; sermons seldom hinder us from pursuing our self-interest as we construe it. Hence the need to be a little more enlightened about what our self-interest is.[14]

Had this stigma of preachiness not existed (Is Christian moralistic zeal to blame here?), and had popular ecoliterature used the labels "virtue" and "vice" as freely as it explores new "attitudes," then surely academics would have noticed sooner what riches lie within this discourse.

At any rate, it may be perplexing that the riches of ecological virtue language are not better known, but it is even more perplexing to contemplate how this language emerged, with such a profusion of variations, in such a short time, and despite the stigma attached to the *idea* of virtue language. Although I have no certain answer, I would like to offer some conjectures on this point.

One reason for the flourishing of ecological virtue language may simply be that this is how, at least in the West,[15] we tend to talk when we are worried and would like to see things change. Virtue discourse, in Western languages, connotes a combination of intention and action that shows a seriousness about thorough and lasting change. It is the discourse we use when we are willing to make commitments and to express these publicly. (Think, for example, of the role of terms like "courage" and "loyalty" during wartime, or terms like "responsibility" and "caring" in the family values debate.) Since ecologically minded people tend to perceive the current environmental crisis as extremely serious, it should not be surprising if they resort to those traditional linguistic constructions that connote active commitment, even if they do not call this language by its traditional name.

Second, I also believe that Lynn White's influential article, "The Historical Roots of Our Ecologic Crisis," has stimulated the use of virtue language among ecologically minded people. White, we may recall, argues that orthodox Christian doctrine has inspired harmful and arrogant "attitudes towards nature." As an antidote to this tradition of domination, and as the first step in a truly thoroughgoing social transformation, White recommends following St. Francis's "belief in the virtue of humility."[16] In response to White's article, a heated debate ensued about Christianity's degree of guilt in contributing to the environmental crisis. Many Christians sought to rectify what they perceived as a misrepresentation of the concept of stewardship. Others, Christians as well as non-Christians, accepted the criticism and went on to explore alternatives. But critics and supporters in this great debate had one thing in common: Since White's thesis hinges on the necessity of a change of *attitude*, they focused their investigations and energies on *attitudes*. This development, I believe, significantly stimulated the use of ecological virtue language—and it may also have stimulated the use of the term "attitude" as a substitute for both "virtue" and "vice."

Still, neither of these conjectures seems to get to the heart of the matter. Linguistic traditions and fashionable topics of debate remain only external incentives for the development of a moral language. They may trigger trends, but they cannot fully account for the kind of proliferation and creativity we find in ecological virtue language. Such flourishing must ultimately be based on a good fit between a moral language and people's ways of experiencing and seeing the world.

Virtue language, in many ways, fits well with efforts to live out an ecological worldview.[17] The cultivation of virtues allows and encourages us to integrate emotions, thoughts, and actions. Thus it fits with the ideal of personal wholeness that many ecologically minded people espouse. The cultivation of virtues depends on narratives, vision, and the power of examples. It does not require external commands or force. This fits with the preference of many ecologically minded people for change through conviction rather than coercion. And, as the many synonyms and closely related concepts in Appendix A illustrate, the cultivation of virtues provides rich expression to many different facets of human experience. This fits with an ecological appreciation for diversity. These may be some of the more substantive reasons explaining why ecologically minded people have discovered the language of virtue.

Whatever the reasons, however, the fact remains that ecological virtue language has emerged and is rapidly growing. In the remainder of this chapter

I will outline five features that characterize dirty virtue language. Obviously, the picture that will emerge is an interpretation, a vision that carries the stamp of my own interests and way of perceiving. Yet I like to think that the picture is not completely arbitrary, that others who pay attention to this emerging ecological virtue language would recognize similar features and patterns. (If not, at least we will have material for a good discussion!)

An Integral Discourse

Ecological virtue language represents a distinct moral discourse with an internal unity and logic—what I will call an integral discourse. This does not mean, of course, that it is entirely separate from other branches of moral discourse, particularly not other virtue discourses. The best way to understand ecological virtue language as an integral discourse is to see it as analogous to a bioregion, which has an integrity and unique character, but which connects with other bioregions and participates in the larger cycles of the biosphere.

What makes ecological virtue language an integral discourse? Not, as one might expect, the presence of a single undergirding ecological worldview, for we find quite a variety of ecological worldviews (see chapter 2). Rather, I would say that the integrity of ecological virtue language derives from its unique *practical* base in the environmental movement, a rootedness that is reflected in a particular moral vocabulary and its conventions. To describe this vocabulary and its accompanying conventions is the aim of this book, a task that cannot be summarized in a few paragraphs.

Intuitively, however, it is easy to grasp the distinctiveness and inner logic of ecological virtue language when one compares it to a moral discourse that has sprouted from a rather different practical base. Take, for example, the virtue language generated by Western corporate cultures. If one needs to make a profit in business, it generally helps to be "tough," "efficient," and "shrewd."[18] It may also help to be caring and trustworthy, but usually these traits are redefined as being, in fact, tough—that is, they produce "hard" results.[19] Although the same combination of traits may aid in the achievement of environmental goals, ecologically minded people generally do not speak this way. First on their lips is not toughness but humility and sensitivity, not efficiency but sustainability, not shrewdness but wisdom. And caring is a basic way of being in the web of relations, not primarily a means to an end. Clearly, business practice and the practice of dedication to ecological well-being generate distinctive ways of using virtue language, each calibrated to the unique goals of the underlying practice.[20]

At this point an objection may well be raised. Can one accurately characterize the environmental movement as a single practical base? Are there not so many internal groups, each with their own approach, often disagreeing with each other, that the idea of a single practice giving rise to an integral moral discourse is no more than an artificial construct? One way to address this objection is to argue that, when push comes to shove, different environmental groups find themselves fighting for a common goal. In his book *Toward Unity Among Environmentalists*, Bryan Norton defends the thesis that anthropocentrists and nonanthropocentrists converge at the practical level of environmental management. Norton bases his claim on the observation that "no long-term human values can be protected without protecting the context in which they evolved."[21] One also hears the related view that seemingly incompatible approaches, such as seeking legislative change and using illegal tactics of radical resistance, complement each other strategically: The radical vanguard tends to increase the popularity of the moderate reformers, thus speeding up the process of gaining popular support for environmental causes.

Yet not everyone is comfortable with such irenic interpretations of the practical scene. For example, some nonanthropocentrists are not convinced that anthropocentrists go far enough in their efforts at species protection.[22] And some ecofeminists resist the idea that deep ecology can adequately subsume their practical concerns and the concerns of other justice-oriented ecologists.[23] As Don Marietta points out, the emphasis on differences may be a matter of principle:

> Insistence on a unifying principle is seen [by some feminists] as causing the rejection of the voices of oppressed people. There is no clear denial of the possibility of there being a unifying principle yet to be discovered, but there is strong resistance to making this search the goal of ethical inquiry.[24]

I very much resonate with these concerns. Therefore, I do not want to belabor too much the question of whether ecological virtue language is an integral discourse rooted in a distinct practice. Yet I do think that the sheer existence of debates and practical disagreements between ecologically minded people shows that there is a degree of common ground, an ecological culture, *which makes disagreement meaningful*. This point is underscored by the practical fact that these debates take place in shared forums such as journals and conferences.

As an integral discourse, ecological virtue language has a distinct vocabulary and operates according to a discrete set of conventions, just as a bioregion displays a distinct flora and fauna bound together by custom-made patterns of interaction. Thus certain virtues and vices occur across the board in environmental literature, typifying as it were the moral landscape.[25] For example, "respect for nature" is a commonly used expression. Different authors might mean somewhat different things by it, of course, but the expression seems to be universally accepted and used. The same applies to the virtues of adaptability, benevolence, care, compassion or solidarity, connectedness, creativity, cooperation, fostering, friendship, frugality, gratitude, healing, hope, inclusivity, joy, justice, moderation or restraint, openness, passionateness, perseverance, realism, self-examination, sensuousness, sharing, spontaneity, vulnerability, wisdom, and wonder. As for vices, ecological discourse is typified by terms such as anthropocentrism, arrogance, carelessness, competitiveness, consumerism or greed, contempt, cruelty, denial, despair, domination or mastery, dualistic thinking, elitism, exploitation, (inexcusable) ignorance or thoughtlessness, indifference, insensitivity, manipulation, pride, otherworldliness, reductionism, romanticism, and wastefulness.

Clearly, not all of these terms are unique to ecological virtue discourse, just as not all the plants and animals that typify a bioregion are unique to it. It is the particular combination of terms, as well as the distinctive ways in which these terms are interpreted and related to each other, that makes for a unique, integral moral discourse.

A Diverse Discourse

Having highlighted the integral character of ecological virtue language, I also want to highlight its internal diversity. By this I do not so much mean to say that we find here a highly diversified set of virtues and vices (though that is the case, as one can see from Appendix A), but rather that no two ecologically minded people speak and write in exactly the same way. To some extent the presence of such diversity is unremarkable, because variation occurs between the speakers of any language. Yet differences in virtue discourse can betray differences in the worldviews, backgrounds, and characters of ecowriters. Such variations are significant and must be highlighted if we are to honor and safeguard the inclusivity of ecological moral discourse.

For one thing, different authors employ different configurations of virtues and vices. One could say that their *catalogues* differ, as long as one remembers that these catalogues are never completely spelled out. Occasionally an author will even go so far as to exclude explicitly a virtue or vice that others deliberately include. The next two chapters, which contain catalogues based on the work of Murray Bookchin and Thomas Berry, illustrate the level of variety on this point.

Diversity also shows up in the form of different emphases. What may be a particularly important attitude to one author may be of little interest or even objectionable to another. To use a traditional expression: Different authors may identify different cardinal virtues. For example, poet and essayist Linda Hogan writes: "Caretaking is the utmost spiritual and physical responsibility of our time, and perhaps that stewardship is finally our place in the web of life, our work, the solution to the mystery of what we are."[26] Yet ecofeminist Elizabeth Dodson Gray finds such a view of stewardly care presumptuous, since "we still in our gut feel we're above and we do not have to fit in." Gray suggests that the key attitude we must cultivate is rather attunement:

> That means you've got to listen. It's like what I consider good parenting to be—cybernetic. It means being guided by information feedback: you do one thing tentatively and you wait to see what happens. You monitor it very carefully (which we do not bother to do). If it looks like it's not doing well, you pull back and you change your behavior, trying something else. Attunement means listening, it means adapting yourself. It means fitting in. And it will never be done by us as a culture unless we, as a culture, realize we are not above.[27]

Not surprisingly, as these examples illustrate, disagreements about key moral attitudes tend to parallel more basic differences within ecological thought, such as the difference between hierarchical and radically nonanthropocentric theories.

Generally, diversity in the area of cardinal virtues is not a topic of explicit debate. Yet one can identify variations between authors by comparing which virtues receive the most emphasis or turn up most frequently in their works. For example, the most-discussed virtue in Wendell Berry's *A Continuous Harmony* is discipline: We need discipline in focusing on details on the farm, in mending our ways for the sake of conservation, in carefully determining proper means to a desired end, in exercising self-criticism, and so on.[28] By contrast, discipline only turns up once (as the discipline of prayer needed for ecological sensitivity) in Sean McDonagh's *To Care for the Earth*, despite the author's Roman Catholic heritage. This is not to say that these authors entirely diverge on what are the most crucial virtues in an ecological age. Both Berry and McDonagh often stress care, love, and respect or reverence. Still, the example of discipline reminds us that in speaking about dirty virtue language, we must remember that what is cardinal to one ecologically minded person may be minor to another.

Similarly, what may be a particularly worrisome vice to one author may seem like a small or misplaced concern to another. To use a traditional expression again, different authors may identify different capital vices. *Hubris* (pride), for example, is an important vice in James Nash's *Loving Nature*. It expresses itself in human attempts to "play God" in relation to nature.[29] In ecofeminist literature, one rarely finds this orthodox theological term, whereas psychological and sociological terms like "alienation," "control," "domination," and "denial" resound like a familiar refrain. In a few extreme cases of variation, what is a most serious vice to one author represents a great good to another. Ecofeminists Marti Kheel and Chaia Heller, for example, argue that calls to "save" and "protect" nature are very harmful, since they diminish and objectify the nonhuman world. Rather than such hierarchical, heroic attitudes, we need holistic virtues of allied resistance and attentive care.[30] Yet saving and protecting are exactly the watchwords of popular environmentalism. "Save the Whales!" and "Protect the Forests!" are battle cries we have all heard (or seen on bumper stickers).

Diversity also shows up when particular virtues and vices are interpreted differently. John F. Haught, for example, highlights variations in the meanings of environmental abuse and care for nature. He correlates the variations with different theological visions:

> In the sacramental view we condemn environmental abuse because it is a sacrilege. But in the eschatological perspective the sin of environmental abuse is one of despair. To destroy nature is to turn away from a promise. What makes nature deserve our care is not that it is divine but that it is pregnant with a mysterious future.[31]

To find such an explicit articulation of diverse interpretations within ecological virtue language is actually rare. Most variations still remain to be uncovered. The question of whether there are recognizable patterns of variety, patterns that might be correlated with distinct branches of the environmental movement, contains much unexplored promise.

Finally, I want to highlight the diverse ways in which virtues and vices can be causally linked. Christian animal rights activist Andrew Linzey, for example, attributes our dominant abuse of animals to spiritual immaturity, which he interprets as a lack of Christian theocentrism.[32] Yet ecofeminist animal rights activist Lori Gruen sees Judeo-Christian religious belief exactly as *causing* our instrumentalist use of animals, since it would encourage the view of animals (and women) as "others" to be used, and even sacrificed.[33] On the virtue side, mainstream Christian authors typically argue that respect for nature follows from faith in a transcendent Creator.[34] By contrast, deep ecologists trace our capacity for respect to the virtue of identification with the larger Self of the natural world. Causal relationships between vices and between virtues have long been of interest to ethicists and theologians, since these relationships affect our ability to change harmful habits and cultivate helpful ones. The diversity of relationships that emerges from ecoliterature calls for renewed reflection on this topic.

Differences between catalogues, cardinal virtues, capital vices, and causal relationships indicate that ecological virtue discourse is a lived language that has neither been logically constructed nor psychologically tested. In its current multifaceted form, we can treasure it as a discursive correlate of the diversity we find in the natural world, a diversity that carries with it the potential for creativity and regeneration. Still, a degree of systematic reflection on this lived language is needed to help distinguish between mere messiness and variations that are pregnant with significance and moral promise.

A Dialectical Discourse

The diverse ways in which different ecowriters use virtue discourse serve as a reminder that we must not too hastily draw general conclusions about this moral language. Nevertheless, we may notice patterns that, with appropriate reservation, help us understand the structure of ecological virtue discourse. In this section I will highlight a pattern of logical and psychological tensions. To the extent that these tensions are deliberate and produce desirable results that cannot be achieved otherwise, ecological virtue language can be called a dialectical discourse.

Not all tensions are deliberate, however. Some represent the kinds of inconsistencies one can expect to find in an ethic that is still in its infancy. For example, I am puzzled when I read the following:

> Love is respect's strongest foundation, although it is often difficult to uncover because of our own desires and fears. Nevertheless, seeing value in another regardless of his/her/its utility to us is a crucial first step towards discovering that love. Without respect we cannot love.[35]

In this passage I first read that love is the foundation for respect, but then I learn that the opposite is the case: Respect is a necessary condition for love. Such convoluted thinking (unless it is an intentional exploration of circularity) can only confuse those who are genuinely interested in cultivating these virtues.[36] Inconsistencies of this kind will need to be straightened out in the interest of moral clarity.

Other tensions within ecological virtue discourse seem to be deliberately maintained in the interest of securing a delicate balance of characteristics. Take, for example, the virtues of earthiness and attunement, both widely valued in ecoliterature. Each points us in a different direction. Earthiness demands practicality, a no-nonsense attitude, even a degree of bluntness. All of this is necessary if we are serious about offering down-to-earth solutions to urgent problems. Attunement, however, requires an opening up of all our senses to the greatest degree of sensibility. It demands gentleness, a dwelling on details, and personal vulnerability. One could say that as virtues, earthiness and attunement want different things for our hearts. Earthiness encourages a certain hardening of our hearts for the sake of justice; attunement encourages a softening of our hearts for the sake of love. Yet both are necessary to keep our hearts in the right place. If earthiness is not balanced by a tuning in to the other, it may turn into a crude form of instrumental reasoning that inflicts unnecessary harm. If attunement is not balanced by earthy commitment to justice, it may turn into a form of nature-romanticism that provides an escape, but not an ethic (and hence confirmation of the status quo). The tension between these ecological virtues is needed to avoid extremes that could lead ecologically minded people in dangerous directions.

Thus, whereas some tensions reflect the infancy of ecological virtue discourse and must be straightened out in due time, others reflect a dialectical richness that ensures flexibility and prevents harmful extremes. This feature challenges us to take another look at the Aristotelian image of a virtue as a mean between extremes of excess and deficiency. While ecological virtue discourse seems geared to avoid extremes as well, it does so in an innovatively relational way. We need not always rely on individual virtues to achieve balance. We can rely on a *network of virtue relations*. This view of the life of virtue nicely corresponds with the general emphasis in ecoliterature on relational or holistic modes of thinking, acting, and being.

A Dynamic Discourse

As a diverse and dialectical discourse, ecological virtue language is obviously hard to pin down. The fact that it is also a dynamic discourse further underscores the futility of such an effort. Ecological virtue language is *emerging*; its shape is not fixed, nor is it likely to become fixed in the near future. To some extent this dynamism can be attributed to the newness of the discourse, which has not yet settled or come to full fruition. But dynamism also belongs to ecological virtue discourse as an integral feature, since this discourse is connected with various evolving realities. Here we discover another recognizable pattern.

One connection involves the natural sciences. New insights from the natural sciences require constant reexamination of basic moral assumptions. For instance, recent discoveries about the importance of symbiosis in biological evolution have given rise to speculation about a possible natural basis for relational virtues such as cooperation, benevolence, and care.[37] Such a natural basis would imply that the life of virtue is less a matter of exercising control (a notion that had drawn support from the prior emphasis on competition in biology) and more a matter of fostering spontaneity. In fact, in a significant move away from established control-centered traditions, ecologically minded authors rarely portray the cultivation of virtues in terms of mastering unruly impulses.[38] Searching for a moral language that suits new biological (and psychological) emphases, they stress the need for a vulnerability that gives basic instincts a chance to be expressed. Their search in this key area, which cuts to the core of the moral life, illustrates how the connection between scientific discourse and ecological virtue language introduces a dynamism with far-reaching consequences.

Another source of dynamism is our changing understanding of the possibilities of human and nonhuman existence on this planet. When we still believed that the Earth could, with the right technology, produce enough food for all people, our moral focus was on technological creativity, redistribution, and development. Virtues like simplicity and responsibility predominated in environmental discourse (this ethic was primarily addressed to rich elites). Now that many of us can no longer support this belief, and now that we are developing a greater interest in the plight of nonhuman life as well, our moral focus is shifting to the challenges of fashioning integrated forms of existence. As a result, virtues like humility, vulnerability, and feeling kinship are becoming increasingly prevalent in ecoliterature. Clearly, such dynamism is necessary if our cultivation of virtues is to be attuned to realistic possibilities in an ecological age. Lack of such attunement could lead us to embrace "virtues" that, despite our good intentions, undercut rather than follow the options that are still open.

Because ecological virtue language has to be dynamic, it is unlikely that this discourse will ever become systematized to the extent that, for example, Thomas Aquinas systematized Christian virtue language. Ecowriters appreciate the ever changing and multifaceted structures of the world, an appreciation that contradicts efforts to build a closed intellectual system with strong claims to ultimacy. In fact, an ecological virtue theory that claimed to be *the* virtue theory would constitute an oxymoron.

It is one thing to value dynamism and quite another to engage in intellectual sloppiness. Insofar as ecological virtue discourse is still a new moral language, there is much room and need for development, for aiming at greater precision of expression, for better understanding conceptual connections and distinctions. As of yet, while the literature contains many virtues and vices, few have been thoroughly interpreted. And conversely, while some traditional virtues and vices receive no mention at all, the reasons for this silence are usually not spelled out, so that the reader is left to wonder whether the omission expresses the author's disapproval or rather the opposite, automatic acceptance. (The virtue of cleanliness, for example, comes to mind here.) Efforts to address these and other instances of incompleteness need not undercut the dynamism of ecological virtue ethics, as long as they flow from a desire to achieve greater moral clarity rather than from a desire to build the ultimate intellectual system.

A Visionary Discourse Without a Social Ethic

By expressing themselves through dirty virtue language, ecologically minded people express an interest in the cultivation and transformation of their own and other people's characters. Yet unlike Aristotle and many of his followers, ecologically minded people are not interested in character development for the sake of achieving personal harmony within an existing social system. Rather, their dirty virtues are tied to a social vision for the future, a vision of ecologically sustainable societies. The change from present social structures to ecological societies is seen as necessary for the survival of diverse life-forms on Earth. And, in the viewpoint of many, the cultivation of appropriate attitudes can foster such change.[39]

Ecological virtue discourse thus derives much of its impetus from a social ideal. Ironically, however, most ecological virtue language does not display the features of a social ethic. Calls to respect nature, to change our dominating attitudes, to be frugal, careful, and wise tend to remain just that: calls. Usually they are not followed by a detailed analysis of how heeding them will bring about the desired social change. Although much ecoliterature does address legal, communal, and institutional changes required to build sustainable societies, virtue discourse is rarely an explicitly integrated part of these specific discussions. Even if a single work contains much virtue language as well as a theory of social change, the two aspects tend to be in different chapters or sections of chapters. Thus we are confronted with the irony that the language of dirty virtues witnesses to a social vision without being connected to a social ethic.

Again, I believe Lynn White's article, "The Historical Roots of Our Ecologic Crisis," was instrumental in setting this pattern of "a social vision without a social ethic." White's thesis that attitudes must change if we are to avert further ecological disaster provides a clear image of ordered social progression, which puts a vice-to-virtue transformation at the start. At the end of the progression we find a vision, inspired by the teachings of St. Francis, of a harmonious, radically inclusive and egalitarian world. Yet White does not indicate what, if anything, beyond the vice-to-virtue transformation would be necessary to get to this better world. His silence on this point was transmitted to the widespread debates that followed the publication of his articles, debates about the core questions of the causes behind and possible solutions to the ecological crisis. Although virtue language flourishes profusely in these debates, social analysis is often thinly represented or totally absent and the listener is expected to simply go along with the assumption that all will pan out as long as we change our basic attitudes (there are exceptions, most notably the work of social ecologists and of many ecofeminists).[40] The end result is the curious phenomenon of a discourse committed to social change without a developed theory of social change.

In my view, this phenomenon calls for critical reflection. I see the current situation as somewhat analogous to the case of the earliest Christians, who were energized by a vision of social transformation through the coming of God's Kingdom, who fervently called for people to practice virtues such as neighborly love, peacefulness, and long-suffering, but who did not develop a social ethic.[41] Hence the early churches did not present a real challenge to reigning powers or to practices such as war and slavery. Yet the history of Christianity also provides a different model. In the nineteenth and early twentieth centuries, the Social Gospel Movement responded to widespread social problems by enriching early Christian eschatological teachings with the insights of contemporary social analysis (at that time especially socialist theory). By so doing the movement was empowered to bring about widespread changes, including legislative changes that ended the exploitation of child labor and limited the workweek to six days. A similar move could now provide more bite to the commitment to social change that characterizes ecological virtue language.[42] In effect, this means that this discourse needs to become more integrated with those sections of environmental literature where we do find the development of a social ethic.[43]

CONCLUSION

Ecological virtue discourse, as a distinct, diverse, dialectical, dynamic, and visionary moral language, deserves more attention than it has received so far. Many riches hide among the dirty virtues, as well as significant challenges to ancient traditions. While these facts may be sufficient to warrant academic interest, there should be a practical benefit as well. Rich moral discourse, rooted in transformative praxis, carries the promise of moral creativity. We are sorely in need of such creativity as we face the many ecological challenges ahead and as we increasingly find ourselves perplexed by the difficulties involved in adapting traditional moral languages to the needs of our age. One more language is one more chance. It would border on irresponsibility not to pay attention to the virtue discourse that emerges from the environmental movement.

In fact, with this new virtue language we may not just be given another chance, but even a *good* chance at achieving moral breakthroughs, at finding fresh ways of looking at problems and dilemmas that are already getting stale (such as the question whether trees or rivers have rights). Virtue language has premodern roots, and although it comes to us sifted through the mazes of modernity, its internal consistency and comprehensibility are not dependent on the worldview that came into power with the scientific and industrial revolutions. Given that many critics see the modern worldview as an important factor in bringing about the ecological crisis, it will be helpful to have access to a form of moral discourse that is not too much in cahoots with this worldview. Even though virtue ethics may have acquired an image of conservatism, a virtue ethic based in the lived discourse of the environmental movement could have surprisingly radical effects.

NOTES

1. Marti Kheel, "From Heroic to Holistic Ethics: The Ecofeminist Challenge," in *Ecofeminism: Women, Animals, Nature*, ed. Greta Gaard (Philadelphia, PA: Temple University Press, 1993), 261.
2. Anthony Weston, *Back to Earth: Tomorrow's Environmentalism* (Philadelphia, PA: Temple University Press, 1994), 58.
3. Lester R. Brown, *Building a Sustainable Society* (New York: W. W. Norton & Company, 1981), 354.
4. Matthew Fox, *Creation Spirituality: Liberating Gifts for the Peoples of the Earth* (San Francisco: HarperSanFrancisco, 1991), 28.
5. Stephanie Lahar, "Roots: Rejoining Natural and Social History," in *Ecofeminism: Women, Animals, and Nature*, ed. Gaard, 111.
6. See Geoffrey B. Frasz, "Environmental Virtue Ethics: A New Direction for Environmental Ethics," *Environmental Ethics* 15 (1993), 259–74; Val Plumwood, *Feminism and the Mastery of Nature* (New York: Routledge, 1993).
7. Only a handful of ecological writers have deliberately called for, developed, or explored virtue-based approaches, often in response to what are perceived as more mainstream approaches. See Thomas E. Hill, "Ideals of Human Excellence and Preserving Natural Environments," *Environmental Ethics* 5 (Fall 1983), 211–24; Eugene C. Hargrove, "The Role of Rules in Ethical Decision Making," *Inquiry* 28 (1985), 3–42; Paul W. Taylor, *Respect for Nature* (Princeton: Princeton University Press, 1986), esp. 198–218; Jay B. McDaniel, *Of God and Pelicans* (Louisville, KY: Westminster/John Knox Press, 1989), esp. 73–4; Plumwood, *Feminism*; Frasz, "Environmental Virtue Ethics," 259–74; Ron Erickson, "Comment on Environmental Virtue Ethics," *Environmental Ethics* 16 (1994), 334–6; James A. Nash, *Loving Nature* (Nashville, TN: Abingdon Press, 1991); John Patterson, "Maori Environmental Virtues," *Environmental Ethics* 16 (1994), 397–409; Laura Westra, *An Environmental Proposal for Ethics: The Principle of Integrity* (Lanham: Rowman & Littlefield, 1994); Jamie Schaefer, "Ethical Implications of Applying Aquinas's Notions of the Unity and Diversity of Creation to Human Functioning in Ecosystems" (Marquette University: Ph.D dissertation, 1994); Pamela A. Smith, "Aquinas and Today's Environmental Ethics: An Exploration of How the Vision and the Virtue Ethic of 'Ecothomism' Might Inform a Viable Eco-Ethic" (Dusquesne University: Ph.D. dissertation, 1995); Deborah D. Blake, "Toward a Sustainable Ethic: Virtue and the Environment," in "*And God Saw That It Was Good*," eds. Drew Christiansen, S.J., and Walter Grazer (Washington, DC: United States Catholic Conference, 1996), 197–210; and Michael S. Northcott, *The Environment and Christian Ethics* (Cambridge, UK: Cambridge University Press, 1996), esp. 121–3, 314–7 (I have not been able to consider the 1996 publications in my book, due to the logistics of publishing).
8. Plumwood, *Feminism*, 173.
9. David G. Hallman, *A Place in Creation: Ecological Visions in Science, Religion, and Economics* (Toronto: United Church Publishing House, 1992), 27.
10. Sean McDonagh, *To Care for the Earth: A Call to a New Theology* (Santa Fe, NM: Bear & Company, 1986), 49.
11. Thomas Berry, *The Dream of the Earth* (San Francisco: Sierra Club Books, 1988), 5.
12. Bill Devall, *Simple in Means, Rich in Ends: Practicing Deep Ecology* (Salt Lake City, UT: Gibbs Smith, 1988), 186.
13. Joanna Macy, "Faith, Power, and Ecology," in *This Sacred Earth: Religion, Nature, Environment*, ed. Roger S. Gottlieb (New York: Routledge, 1996), 417, 421.

14. Joanna Macy, "Awakening to the Ecological Self," in *Healing the Wounds: The Promise of Ecofeminism*, ed. Judith Plant (Philadelphia, PA: New Society, 1989), 209. In this passage Macy elaborates on Arne Naess's criticism of the traditional notion of altruism.
15. Ecological virtue language also appears in non-Western texts. One thinks, for example, of the role of *ahimsa* and benevolence in Buddhist environmental literature. However, my expertise does not extend to these areas. I do hope that in the future comparative studies will be carried out.
16. Lynn White, Jr., "The Historical Roots of Our Ecologic Crisis," *Science* 155 (10 March 1967), 1206, 1207.
17. For a description of the cluster of worldviews that can be designated as "ecological," see chapter 2.
18. On toughness as a business virtue, see Robert Solomon, *Ethics and Excellence: Cooperation and Integrity in Business* (New York: Oxford University Press, 1992), 213–6.
19. See Louke M. van Wensveen, "Is Toughness a Business Virtue?" *The International Journal of Applied Philosophy* (Winter/Spring 1995), 15–25.
20. I do not mean to imply here that either the corporate world or the environmental movement has a single goal or set of goals that is uniformly interpreted and accepted. I am thinking rather of two nodes of conglomerate goal-oriented behavior where each aggregate of goals is sufficiently distinctive to yield an identifiable practice and matching moral discourse.
21. Bryan Norton, *Toward Unity Among Environmentalists* (New York: Oxford University Press, 1991), 240.
22. Brian K. Stevenson, for example, criticizes Norton's model for not accommodating nonanthropocentrist requirements for species protection ("Contextualism and Norton's Convergence Hypothesis," *Environmental Ethics* 17 [1995], 135–50).
23. See Deborah Slicer, "Is There an Ecofeminism—Deep Ecology 'Debate'?" *Environmental Ethics* 17 (1995), 151–69.
24. Don E. Marietta, Jr., "Pluralism in Environmental Ethics," *Topoi* 12 (1993), 70.
25. I make these claims very hesitatingly, since I am wary of attempts to seek or construct a unified environmental ethic. My observation that ecological virtue language (still) forms a single discourse represents what I see as the most fitting interpretation of my findings; it does not represent a program. My own position of pluralism resembles Don Marietta's statement above.
26. Linda Hogan, "The Kill Hole," in *This Sacred Earth*, ed. Gottlieb, 40.
27. Elizabeth Dodson Gray, "A Critique of Dominion Theology," in *For Creation's Sake: Preaching, Ecology, and Justice*, ed. Dieter Hessel (Philadelphia: Geneva Press, 1985), 80.
28. Wendell Berry, *A Continuous Harmony: Essays Cultural and Agricultural* (San Diego: Harcourt Brace Jovanovich, 1970, 1972), 53, 81, 86, 108.
29. Nash, *Loving Nature*, 104.
30. Chaia Heller, "For the Love of Nature: Ecology and the Cult of the Romantic," in *Ecofeminism*, ed. Gaard, 235; and Kheel, "From Heroic to Holistic Ethics," ibid., 248.
31. John F. Haught, *The Promise of Nature* (New York: Paulist Press, 1993), 110.
32. Andrew Linzey, *Christianity and the Rights of Animals* (New York: Crossroad, 1987), 15ff.
33. Lori Gruen, "Dismantling Oppression: An Analysis of the Connection Between Women and Animals," in *Ecofeminism: Women, Animals, Nature*, ed. Gaard, 64.

34. See, for example, Nash, *Loving Nature*, 100.
35. Gus diZerega, "Individuality, Human and Natural Communities, and the Foundation of Ethics," *Environmental Ethics* 17 (1995), 37.
36. There may be instances, of course, in which two or more virtues need to be cultivated simultaneously because they depend on each other. Perhaps the author of the above passage would say this is the case for love and respect also. Yet in that case it is important to clarify how such simultaneous cultivation can be achieved.
37. Many ecological authors, especially ecofeminists, in fact already treat the natural basis of these virtues as a given. Cf. "Our children need to be sustained in their natural empathy with and compassion for animals" ("Liberating Life: A Report to the World Council of Churches," in *Liberating Life: Contemporary Approaches to Ecological Theology*, eds. Charles Birch, William Eakin, and Jay B. McDaniel [Maryknoll, NY: Orbis, 1990], 287).
38. For a critique of "ethics as restraint," see Kheel, "From Heroic to Holistic Ethics."
39. Not all hold this view, however. Christopher Stone, for example, writes: "If the world is going to be changed, it is going to be changed by people who can get past talking up a reform of the human spirit and lessons to be learned from Buddhism, and put in the effort required to understand fisheries quotas, pollution taxes, trade barriers, and what the International Court of Justice can and cannot do" (*The Gnat Is Older Than Man: Global Environment and Human Agenda* [Princeton: Princeton University Press, 1993], xiv).
40. Many ecofeminists draw on general feminist social theory. Analytical concepts such as patriarchy, oppression, and ideology play an important role in their work.
41. See J. Philip Wogaman, *Christian Ethics: A Historical Introduction* (Louisville, KY: Westminster/John Knox Press, 1993), 30.
42. Such a move would also benefit the development of virtue theory in general, much of which fails to deal with larger questions of social change.
43. One promising step in this direction can be found in Max Oelschlaeger's book *Caring for Creation*. Oelschlaeger sees the virtue of caring for creation as a metaphor embedded in various religious narratives. Such narratives are crucial for social change in democratic societies, he argues, since they inspire discourse on the public good and they challenge the idea that the state as such is already moral. Although social change requires more than religious narratives (i.e., "politically coordinated, scientifically informed, and institutionally empowered actions that are consistent with our basic democratic freedoms and diversity of ultimate commitments"), Oelschlaeger's narrative ethic explicitly integrates the dynamics of social change with the cultivation of virtue (*Caring for Creation: An Ecumenical Approach to the Environmental Crisis* [New Haven, CT: Yale University Press, 1994], 10, 76–7).

    Elsewhere, Oelschlaeger (with Michael Bruner) also argues for the importance of rhetoric in bringing about social change ("Rhetoric, Environmentalism, and Environmental Ethics," *Environmental Ethics* 16 [1994], 377–96). Since virtue discourse is highly suitable for rhetorical purposes, this argument could also be used to construct a theory of the role of virtue in social transformation.

Chapter 2 Two

SOMETHING OLD, SOMETHING NEW

AS I PLUNGED THE depths of the emerging language of dirty virtues, one question continued to preoccupy me: How new is this discourse really? When we seek to understand something, it helps if we can place it in relation to what is already familiar to us. Although it was useful to discover just how diverse, dialectical, dynamic, and visionary ecological virtue language is (for one thing, it cured me of the idea that I might construct a unified ecological virtue theory), the language only really began to take shape for me when I placed it in the context of the Aristotelian-Thomistic virtue tradition, the tradition with which I am most familiar. Through comparative interpretation I became able to appreciate the "roots and wings" of ecological virtue language: its connections with the past and its promises for the future. Thus the dirty virtues became part of a larger story, a story that adds to their comprehensibility and significance.

In this chapter I take the question of the newness of ecological virtue language as my guide—that is, its newness compared to a Western virtue tradition with a history of popularity. As we shall see, it makes the most sense to speak of "similarities within differences and differences within similarities."[1] Dirty virtues are not entirely new. But then they are not entirely old hat either, which is exactly what one would expect from a dynamic moral language.

AN INNOVATIVE DISCOURSE

The virtues and vices that belong to a moral discourse cannot be fully understood as isolated entities. They derive their meaning in relation to each other and in relation to various beliefs and conventions that make up a worldview (or, as in the case of ecological discourse, a cluster of associated worldviews). Hence, in order to assess the innovative character of ecological virtue language, we must first pay attention to the beliefs and conventions that provide the glue[2] of this language. Only after we have compared this glue with more traditional glues can we move to a comparison at the level of specific virtues and vices. (Readers who are well acquainted with ecological thought may choose to skip the first half of the following subsection.)

WORLDVIEW SHIFTS

The cluster of worldviews that is emerging from ecoliterature differs significantly from dominant Western traditions of thought. The majority of ecologically minded authors at some point or another criticize traditional ways of dividing the world into separate and ostensibly unrelated, or opposing and hierarchically ordered parts. As an alternative, they usually suggest one of various forms of holism. I will briefly outline these widespread critiques as well as the proposed alternatives with an eye to their relevance for virtue ethics.

The view that the world consists of billiard-ball relationships, a view that gained prominence in modern history through the influence of Newtonian physics, can be called *atomism*.[3] According to ecowriters, such a view does not accord with the interrelatedness of reality, as attested to by many recent discoveries in physics and biology.[4] Jewish and Christian authors add that it conflicts with their understanding of the created world as a fundamentally relational world. As Douglas Bowman, a Protestant environmentalist, observes, "With the breaking up of our outlook upon the world, we break and rend asunder the biosphere, which becomes itself Humpty Dumpty fallen."[5] As a result of atomism, human treatment of the environment lacks sensitivity and foresight.[6] Our actions are as compartmentalized as our worldview.

Atomism may take on the form of dualism, a way of perceiving the separated parts of the world in terms of binary distinctions. Well-known traditional distinctions include: matter/spirit, body/soul, passions/reason, natural/supernatural, world/God, nature/culture, and female/male. As many ecologically minded authors point out, such distinctions, when used as rigid interpretive grids, lead to serious misrepresentations of the world. After all, most things belong to some continuum that, when looked at closely, does not permit an easy drawing of dividing lines. What is worse, dualistic thinking has invited an either/or mentality, as if one had to choose one part over the other. This has, in turn, led to oppositional thinking, which has found ultimate expression in violent practices. As many ecologically minded authors point out, in these practices nature has usually lost to culture, female to male, and body to spirit.

Oppositional thinking often implies a hierarchy. The term "hierarchical dualism" is used to describe ways of looking at the world that not only radically separate its parts, but that also systematically order them in terms of degrees of value and appropriate levels of power. A typical example of hierarchical dualism is the view that humans are not only completely different from the nonhuman world, but also inherently superior and destined to be in charge. Although the emphasis on radical human uniqueness predates Christianity (consider, for example, Stoic rationalism), the theological tradition that presents the human as the crown of creation, made in the image of God, has been particularly influential in sustaining such a view in the West. This tradition has provided a rationale for why human beings deserve to be treated with respect and why they have certain rights.[7] But as ecowriters point out, a problematic side effect of such teaching is the accompanying sense of distance from and even contempt for the nonhuman world.[8] This is especially dangerous nowadays, given the awesome destructive capabilities of modern industrialized societies.[9]

The idea that humans are superior has also encouraged anthropocentrism, a kind of moral myopia that makes human well-being the measure of all value.[10] (Many ecofeminists prefer to speak of "androcentrism," arguing that de facto it is *men* who have been the measure of all value.) Again, Christian tradition has been instrumental in sustaining anthropocentric (and androcentric) value systems in the West. For ecofeminist theologian Sallie McFague, it's all summed up in Augustine's statement, "God and the soul, nothing more, nothing at all."[11] From an ecological perspective, the problem with such anthropocentrism is that it reduces the nonhuman world to a stage on which the human (or even just the male) drama takes place.[12] This creates a setting for treating the nonhuman world as a throwaway object.

The alternatives that ecologically minded authors propose for these various types of atomistic and hierarchical thinking usually involve some form of holism. In a general sense, one can define holism as a paradigm in which everything exists in relation to everything else. As McFague points out, this paradigm is marked by a softened sense of divisions. Traditional divisions are not abolished, but rather reinterpreted as representing (God-given, some would say) diversity and otherness.[13] No part of the relational network is inherently devoid of value, or evil. Where evil exists, it is those forces or circumstances that harm the sustainable interaction of the whole and its parts.

There is quite some variety among these holistic alternatives (which is one reason why one cannot speak of a single ecological worldview; certain authors even object to the term "holism" altogether). Some environmentalists propose a "strong" form of holism, in which all things participate in a single ontological unity, a deeper reality they tend to describe and approach with mystical language. Others prefer a "weak" form, in which everything has a common origin and is intricately connected in a functional network, without the assumption of any deeper unifying principle.[14] Both strong and weak holism can be either hierarchical or egalitarian. To illustrate the latter distinction, I will return to the question of the relationship between humans and the nonhuman world.

Environmentalists often highlight the continuities between human beings and the nonhuman world. As Thomas Berry puts it, "we need a way of designating the earth-human world in its continuity and identity rather than in its discontinuity and difference."[15] Appreciating such continuity involves not only intellectual understanding, but an entirely new attitude, marked by a sense of communion with the Earth and a great loving care for all our brother and sister creatures. When it comes to describing the character of the human/nonhuman continuum, however, opinions are divided.

Some environmentalists still see the continuum as consisting of hierarchically ordered relationships.[16] This interpretation is particularly popular among mainline Christian environmentalists, for it allows them to reappropriate traditional theological language. Australian theologian Geoffrey R. Lilburne, for example, describes the place of humans in creation as follows:

> To be in the image of God is not to arrogate to humanity a place of independence from the rest of creation. Rather it is to assign humans a determinate place in the order of being, as lower than God but higher than other creatures.[17]

Thus the notion of a natural order, traditionally of interest to theologians as a way to distinguish between various levels of creation, has now become a way to describe the coherence and connectedness of creation.

Other ecologically minded authors prefer to describe the human/nonhuman continuum in terms of physical similarities and interdependence, without attaching hierarchical notions.[18] Ecofeminist Elizabeth Dodson Gray graphically illustrates this way of seeing the world:

> In the miracle of photosynthesis, plants, when the sun shines, breathe out oxygen to us. We in turn breathe out carbon dioxide back to them. Once again, it is a never-ending cycle of cooperative community or relationship. We and the plants are totally dependent on each other, each to breathe out so the other can breathe in.

This view of interdependence leaves room for human uniqueness, but it does not allow for human superiority. Creation is thus radically egalitarian. Again in Gray's words:

> A bioecological perspective views the human species as unique also, but not necessarily as the most important or the most adept. Certainly we don't have the best ears or the fleetest feet or the keenest sense of smell. We don't even have the largest brains, nor can we echolocate as do the dolphins. And we would be hard-pressed to show that what humans do is as important or as foundational to all life on this planet as what green plants do in photosynthesis.[19]

The differences between hierarchical and egalitarian holism are significant.

Yet in light of long-standing Western traditions, the similarities are more striking. While neither position involves a denial that human beings are unique, both concentrate on how human beings fit in. For example, in Christian circles both groups like to refer to the earthly origins of human beings ("Adam" is translated as "earth creature" or "earthling"). Images of weblike existence abound in both Christian and non-Christian literature.[20] Also, hierarchical holism as well as egalitarian holism tend to issue into calls for humans to feel part and parcel of our ecosystem, and to think in ways akin to the workings of nature.[21]

Overall, the critiques of atomism, dualism, hierarchical dualism, and anthropocentrism/androcentrism as well as the holistic alternatives that characterize environmental literature represent a significant diversion from mainstream Western traditions. This diversion has left its mark on ecological virtue language. In the remainder of this section I will highlight four implications.

First, the critique of atomism raises questions about the common practice of discussing the life of virtue in isolation from its natural context. If there is truth in the holistic claim that everything is connected, then the current deterioration of the natural environment can be expected to affect the moral life as well. Interestingly enough, history already shows that ecological disasters accompanied the decline of many great civilizations.[22] Destruction of the environment, the collapse of communal structures, and the decline of virtues have gone hand in hand in the past. That communal structures and the life of virtue are integrally connected has been pointed out by narrative ethicists. As Stanley Hauerwas puts it, "sharing a worthy adventure" is "soil crucial for the growth of virtue."[23] Now environmentalists are taking this argument one step further when they suggest that good soil—and trees and air and water—itself is also crucial for the growth of virtue.[24] A few lines by Walt Whitman best sum up this growing insight: "Now I see the secret of the making of the best persons. It is to grow in the open air, and to eat and sleep with the earth."[25] In the next two chapters we will see further illustrations of how a connectional way of thinking impacts ecological virtue language.

Second, the critique of dualism calls into question any dualistic opposition of virtues to vices. A dualistic understanding of the relationship between virtues and vices derives from ancient as well as popular assumptions that radically opposed forces of good and evil operate behind the virtues and vices, forces that make the human soul their battlefield. In such a view, virtues are typically associated with spirit or reason, whereas vices primarily operate through matter or flesh.[26] In a dualistic paradigm, virtues also tend to be presented as antidotes to vices: The way to combat a vice is to practice the opposite virtue.

This dualistic interpretation of the life of virtue has long been part of Christian tradition. We find it in the earliest Christian literature, such as the letters of Paul (e.g., Gal. 5:16–24) and the *Didache*, a catechetical document that contrasts "the way of life" to "the way of death."[27] We also find it in the medieval penitentials, which advise that vices be "cured" by applying contrary virtues.[28] The tradition continues today in popular understandings of virtues and vices, as well as in the thought of spiritual leaders such as Pope John Paul II. In his encyclical *Sollicitudo Rei Socialis*, for example, the pope argues that "desire for profit" and "thirst for power" are "attitudes and 'structures of sin' [that] are only conquered—presupposing the help of divine grace—by a *diametrically opposed attitude*," namely, the new virtue of solidarity.[29] In the area of virtue theory this dualistic tradition has recently found expression in Romanus Cessario's *The Moral Virtues and Theological Ethics*. In a discussion on "disordered emotions," for example, Cessario argues:

> Nothing prohibits the believer from exercising an act of authentic personal choosing, for example, to will the good of virtue, even in the presence of strong sense desire for some object opposed to it. Spiritual writers underscore the importance of this truth when they encourage believers to face temptations against the moral virtues by recourse to acts of faith, hope, and charity.[30]

In sum, the dualistic opposition of virtues to vices, as forces pitted against each other, has a long history that takes us right into the present.

The ecological emphasis on holism challenges this ancient tradition. Holism does not require the abolishment of the distinction between virtues and vices, or between good and bad. Rather, it requires the abolishment of the linkage between the virtue/vice distinction and ontological dualisms. Thus ecologically minded authors do not locate the roots of vice in a clearly identifiable evil section of reality, nor in an evil will, but they trace vice primarily to ignorance, pain, deception, or isolation. In other words, the roots of vice are imperfection and brokenness.[31] Vice is tragic rather than evil in origin.[32] Dealing with it is not a matter of imposing an antidote, but of compassionately addressing the causes of human tragedy.[33]

Third, the critique of hierarchical dualism presents a challenge to the popular idea that virtues serve to suppress or correct "natural" impulses or passions. Western traditions tend to link the stirrings of passion with animal behavior.[34] Because animals are seen as lower on the hierarchy of being, the stirrings of passion must be mastered if we are to be truly human. These ideas have a long history in philosophy as well as in theology. In Christian circles, the idea gained momentum when Augustine took the final step in his conversion process, taking as God's special revelation to him the apostle Paul's advice to "spend no more thought on nature and nature's appetites" (Rom.

13:14).[35] The injection of Aristotelianism into Christian tradition, through the work of Thomas Aquinas, did counterbalance the Pauline/Augustinian suspicion of natural desires. Aquinas saw virtues as *directing* "the sensitive appetite," rather than opposing or suppressing it.[36] Yet the Reformation's rejection of Thomism in favor of Pauline and Augustinian theology ensured the continued existence in Christian tradition of the assumption that passions represent fallen nature and need to be suppressed.

A modern version of this long-standing tradition can be found in Gilbert Meilaender's *The Theory and Practice of Virtue*. Due to his indebtedness to both Josef Pieper and Martin Luther, Meilaender stands at a curious juncture with regard to the role of the natural in virtue theory. This accounts for Meilaender's ambiguous treatment of the natural in *The Theory and Practice of Virtue*. On the one hand, Meilaender agrees with Pieper that certain virtues are ours by nature. Meilaender writes, "Among the things which we must approve and affirm when we share in the divine approval of creation are the natural virtues, even our own." On the other hand, this assertion gets overshadowed by repeated references to Luther, who found "nothing profitable" in our natural virtues. The imagery Meilaender uses when he links his own virtue theory with Luther's theology is rather telling in this regard. Even in Luther, argues Meilaender, "there is room for gradual development of character, room for the Christian's assault, aided by grace, upon the fat relentless ego [a phrase Meilaender borrows from Iris Murdoch], room even for the self-mastery which makes progress in virtue possible." Elsewhere, while discussing Plato, Meilaender equates the process of disciplining the "fat relentless ego" with taking seriously "the affective dimension of human behavior." Overall, the reader gets the impression that the life of virtue is a rather violent undertaking, in which grace suppresses rather than perfects nature.[37]

Such forceful mastery, however, may well lead to harmful psychological repression. What is worse, draconian control of "the animal inside us" is likely to arouse contempt for animals and nonhuman nature in general. When virtues become instruments of suppression, people tend to become agents of oppression. Clearly, from an ecological perspective this is worrisome. If the suppression of natural impulses is conceptually linked to the suppression of nonhuman nature, then it is imperative to find another way of experiencing and dealing with our inner stirrings. One finds this latter way implicit in much ecoliterature. It includes a more holistic appreciation for the interwovenness of reason and emotions and even a willingness (sometimes eagerness) to consider passions as important moral guides rather than subverters of the moral life. However, much of this line of thought still needs to be spelled out.[38]

Finally, the critiques of anthropocentrism and androcentrism raise questions about traditional ways of associating the life of virtue exclusively with human flourishing. The aim of human life is happiness, as Aristotle taught, and virtue leads to happiness. This focus on human (Greek male!) happiness or flourishing as the ultimate telos relegates benefits of a person's virtuous actions for the nonhuman world to the status of side effects. In essence, then, the vision underlying this ancient view of the moral life is environmentally (as well as socially) exclusive.

Thomistic virtue theories provide a more complex case of anthropocentrism. On the one hand, Thomas Aquinas and those who stand in his tradition affirm that virtue leads to personal flourishing. Josef Pieper, for example, defines virtues as those excellences that enable a person "to attain the furthest potentialities of his nature."[39] On the other hand, Thomistic tradition also addresses the flourishing of the rest of creation. This comprehensive flourishing is enhanced by the human exercise of virtue. As Jean Porter points out in her book, *The Recovery of Virtue*, all creatures, besides being oriented to self-perfection, are also oriented toward wider goods. One of these goods is the good of the universe as a whole.[40]

This ostensive moral inclusiveness is undercut, however, by the general cosmology as well as the anthropocentric tone of the balance of Thomistic theory. As Jean Porter indicates, Aquinas's universe is ordered in such a manner that the flourishing of higher creatures *necessarily* promotes the good of lower creatures.[41] Thus higher creatures are not continually obliged to remember the plight of lower creatures. If they focus on their own flourishing, then God will take care of the rest through the laws of the universe. In Porter's words, "the good of the universe is God's proper concern, and although we are obliged to conform our wills to God's formally, God is not so unreasonable as to expect us always to will concretely what he wills in his capacity as governor of the whole universe" (I–II.19.10).[42] From an ecological perspective, however, it is both unrealistic and dangerous to assume that the universe is ordered such that human flourishing will always harmoniously benefit the rest of creation. In the current state of the biosphere, much of the previously existing relative balance has been lost. Indeed, the most important threat to the biosphere is the current and projected size of the human population.[43] Even if this entire population were virtuous, other creatures would be crowded out. A holistic, inclusive virtue ethic does not conceal such a troubling fact.

Where a holistic vision of flourishing does lead is more difficult to say at this point. In the next two chapters I will outline two significantly different alternatives. Clearly, much work remains in this area also. Yet one thing is evident: The ecological critique of anthropocentrism undermines any unquestioned reliance on traditional exclusivist teleologies. Ecological virtue discourse does not rely on a narrowly anthropocentric vision of flourishing, but rather, as Thomas Berry puts it, on "the integration of human well-being within the context of the well-being of the natural world."[44] This integrated vision is not entirely new within Western traditions. For example, in the Hebrew Bible as well as in the New Testament one finds references to God's care for nonhuman creation (e.g., Gen. 9:13; Exod. 23:10–12; Matt. 10:29), and to the eschatological renewal of the entire earth (Is. 11:1–9; Rom. 8:22). However, due to the largely anthropocentric focus of orthodox teachings, such inclusive texts have received relatively little attention until recently. It is no coincidence that ecotheologians are recovering this salient aspect of tradition at the same time as they are developing an inclusive virtue discourse.

Clearly, ecological critiques of orthodox Western ideas, as well as the various holistic alternatives, have far-reaching implications for virtue ethics. If we take these critiques and alternatives seriously, we commit ourselves to addressing the natural context of the life of virtue, to a nondualistic understanding of the relationship between virtues and vices, to ways of cultivating virtues that do not, by definition, involve suppression of spontaneous impulses, and to an inclusive vision of flourishing. These commitments, when compared to Western virtue traditions, can be characterized as relatively innovative. Since they function as the glue that connects and penetrates the multiple parts of ecological virtue language, they inevitably lead to innovative interpretations of specific virtues and vices as well.

NEW VIRTUES AND VICES

When I began studying ecological virtue discourse, I was immediately struck by the many creative reinterpretations of traditional virtues and vices. These reinterpretations range from minor adaptations to major extensions or shifts of meaning and even total value reversals. I also noticed *new* virtues and vices: attitudes and actions, deemed either desirable or undesirable, that had hitherto not been the focus of extended moral reflection. In this section I will provide five examples that illustrate the range of innovative interpretation one finds in ecological virtue discourse.[45]

Compassion is an important ecological virtue, since it lays the groundwork for responding to the brokenness of the world. As ecofeminist Rosemary Radford Ruether observes, establishing a new social practice requires "a new sense of urgency about the untenability of present patterns of life and compassionate solidarity with those who are its victims."[46] Ecological interpretations of compassion tend to differ only slightly from traditional interpretations. Environmental authors typically encourage us to extend our compassion beyond human beings, to cultivate "feelings on behalf of the larger whole of which we are a part."[47] Although such calls parallel longstanding concerns of minority traditions such as Franciscan spirituality and animal rights activism, on the whole they represent an adaptation of Western moral traditions, which have typically focused moral concern on the human. Clearly, extending the range of compassion has far-reaching practical implications. It could signify the difference between life and death for many nonhuman beings. Nevertheless, from a conceptual point of view the change is relatively small, since it does not significantly affect the meaning of the virtue of compassion itself. In ecological literature, compassion still connotes the ability to empathize with another's suffering and a corresponding readiness to offer relief.

We do find conceptual innovation when the practice of a virtue is associated with an entirely new goal. Asceticism provides a case in point. Albert Fritsch interprets this virtue as "strengthening will power" in order to work more effectively toward ecojustice. For example, abstaining from meat can be a way to prepare oneself for future difficulties in the fight for ecojustice.[48] Other authors promote vegetarianism in order to spare the lives of other animals and also as a healthier diet. Thus, in an ecological context, asceticism connotes the exercise and discipline necessary *for the flourishing of earthly life*, whether this be healthy personhood, or the restoration of healthy social and ecological relations. This vision of ascetic practice clearly diverts from established traditions, such as the *contemptus mundi* tradition, which associates asceticism with the goal of being freed from material bonds.[49] In fact, insofar as this goal is questionable from an ecological perspective, traditional otherworldly asceticism can only be considered a *semblance* of genuine virtue.

Conceptual innovation can also stem from a new understanding of the moral agent. The virtue of respect illustrates this point. Rather than connoting politeness to superiors ("Respect your parents!"), respect in environmental literature refers to the attention and consideration humans must give to the entire network of relationships in which they find themselves. (Thus ecological respect, like compassion, involves an extension of focus.) At the root of such respect lies not a fear of authority, but rather a marveling at the existence of the other.[50] This new understanding correlates with a new view of the moral agent, who is no longer a peg in a hierarchy of being, but rather a node in a network of relations. As a result, "giving others due consideration" (which one can see as the goal of respect) does not require "looking up," but rather "looking around." Respect becomes a propensity to halt at the mystery of the other and to proceed with care at every encounter. Without such a propensity, the virtue of respect degenerates into the legalistic and insensitive fulfillment of duty, which sooner or later will hurt the very other one claims to respect. Such a degeneration might, in the past, have counted as virtuous (consider the popular opinion that the fulfillment of duty against one's inclination is more commendable than spontaneous fulfillment of duty). Yet in an ecological context, marked by appreciation for inner richness and spontaneity, legalistic respect (like otherworldly asceticism) becomes a mere semblance of virtue.

More radical than conceptual innovation is the complete reversal of moral value. In ecological virtue language, traditional virtues may be treated as vices, and traditional vices may be treated as virtues. Such reversals occur when neither the goal nor the actions associated with a particular virtue or vice are valued as they used to be. Mastery is a good example. Whereas mastery has had positive connotations throughout much of Western history, the concept is now treated with suspicion in ecological literature. Most significantly, mastery of nature, once a hailed objective, has been linked with the present environmental crisis. As numerous authors point out, we have typically destroyed the self-sustaining structures and integrity of the nonhuman world in our attempts to control it. Thus what used to be a virtue has now become a vice.

A final instance of innovation involves the creation of new virtues and vices. For example, the term "attunement," insignificant in the history of Western ethics, has taken on positive moral meaning in ecological discourse. It connotes a continuous monitoring and adapting to the changing situation, needs, moods, and energy levels of other beings as well as systems (such as a bioregion). To a certain extent, attunement is "receptivity of the mind."[51] Yet, as D. Michael Levin points out, this virtue "cannot be taught or learned solely at the level of reflection." He suggests "rooting one's responsiveness in the trans-personal dimension of bodily felt experience."[52] Attunement implies harmonizing with a changing reality.[53] Just as getting the sense of a song requires alertness and anticipation and a feel for rhythm and flow, so does "harmonizing with the dynamics of change in the natural world."[54] Cultivating an attitude of attunement is seen as essential if we are to interact fruitfully in a world we cannot completely control or comprehend.

These examples of focus extension, conceptual innovation, value reversal, and the creation of new virtues and vices illustrate just how innovative ecological virtue discourse is.[55] Much of the innovation can be traced to the critiques and holistic beliefs discussed earlier. Extending compassion to the nonhuman world correlates with the desire to move beyond anthropocentrism. Practicing asceticism in the service of earthly goals correlates with the desire to move beyond the spirit/matter dualism. Understanding respect in an egalitarian way and rejecting the notion of mastery correlate with the desire to move beyond hierarchical thinking. And cultivating attunement as a basic mode of being correlates with the desire to move beyond an atomistic view of the world.

As innovative as these changes are, none are completely unprecedented when put in a more comprehensive historical and social context. In some cases the innovation has been foreshadowed by minority traditions in the West. For example, the rejection of mastery has precedents in the feminist movement. In other cases what counts as innovative in a Western context constitutes established tradition in another culture. Extending compassion to nonhuman beings, for example, is a common Buddhist practice. In all cases, what is new only makes sense in relation to what is old, and thus the old continues to shine through.[56] For example, if one has grown up in a Western culture and wants to practice earthly asceticism, the ghost of its traditional semblance will likely affect one's efforts, perhaps even stimulating an overreaction. But positively also, much that is traditional simply continues to live on in what is new, as we have seen in ecological interpretations of compassion.

In light of these observations I find it most helpful to speak of differences within similarities, or similarities within differences—depending on where the balance lies.[57] No part of ecological virtue language is completely new. This is one reason why, if we truly want to understand this discourse, we must pay attention to its traditional aspects as well.

A Traditional Ethic

Since the overall tenor of ecological virtue language is one of innovation, I was initially surprised to discover to what extent established traditions also live on in this discourse. I came to appreciate how moral creativity, like most creative efforts, is not a *creatio ex nihilo* but more like new growth sprouting from an old stump. No one would confuse the new shoots for the old tree, yet everyone would recognize a connection.

In their attempts to connect ecological virtues and vices with ancient moral wisdom, ecowriters search through many different traditions. Often they are rather eclectic, combining insights from ostensibly quite disparate sources, including all the major world religions, Native American wisdom, witchcraft, Pacific tribal religions, Earth Goddess religion, as well as various schools of philosophy.[58] Such eclecticism highlights the spirit of interreligious, cross-cultural, and ecumenical cooperation that characterizes the environmental movement.[59]

Some argue, though, that Western traditions, especially Christianity, are unsuitable for building an ecological virtue ethic, given how they are implicated in bringing about the environmental crisis.[60] Still, many Western authors have not given up hope in the reformability of their heritage and so they do seek to continue its legacy by searching for familiar roots, even as they are venturing into the largely uncharted territory of ecological ethics. In this section I will focus on links between Western traditions and the structure as well as content of ecological virtue language.

In its formal structure, ecological virtue discourse shows remarkable similarity with the Aristotelian tradition of ethics. The cultivation of ecological virtues is seen as part of a self-realization process that aims at the flourishing of individuals-in-community.[61] (Of course there is a material difference here: For ecowriters this community includes the nonhuman world, whereas for Aristotle and many of his followers, it did not.) Certain attitudes and actions lead to such flourishing, which makes them good and which also reflects well on the characters of the actors involved. These are the virtues, that must be cultivated through committed practice, guided by clear vision as well as understanding of the connections between means and ends (practical wisdom).[62] Where vision and understanding are lacking or misguided, flourishing is diminished and damage may be done. This is where vices enter the picture.

Ecological vices tend to be traced to a lack of true understanding, just as in Aristotle's ethics, vice resulted from following false knowledge.[63] Opinions differ as to what true ecological understanding involves. It might be seeing ourselves as part of the larger Self, or having intimate knowledge of a bioregion, or knowing the Earth to be a living organism. (Again, there is an obvious material difference with Aristotle here.[64]) Opinions also differ as to why people would lack in such understanding. It might be due to a lack of information (which may not be excusable), or as a result of denial or obsessive habits.[65] Yet one thing is clear: Understanding involves more than knowing the facts, since many people who are informed about the environmental crisis do not act on what they know. Rather, the understanding that undergirds virtue and lacks in vice is an integrated kind of knowledge, the kind one assimilates at the deepest level of one's being, the kind that incorporates emotions and, some authors would say, the spiritual dimension of the world. From such knowledge virtue flows spontaneously, just as Aristotle saw the practice of true virtue as a harmonious way of being.[66]

In terms of content, ecological virtue language also shows similarities with established virtue traditions, despite the innovations highlighted in the previous section. Even in the midst of what sounds quite new, one can hear the echos of tradition, especially minority traditions. The attitude of "earthiness" provides a good example. What is meant by it?[67] Earthiness involves the experience of the Earth as one's home, with all its beauty and messiness. Earthy persons fully taste the joys of their home—a sunset, the flavor of chocolate, the hug of a child. Yet they are not romantic dreamers. They will get their hands dirty to do what needs to be done if the home needs cleaning. And they are not perfectionists either. As Albert Fritsch puts it, "activists work best as down-to-earth people, practical enough to know that all undertakings are imperfect."[68] Finally, in order to balance their great love of life with the constant messiness of life, earthy people need a sense of humor. (By tea parlor standards, of course, earthy humor may seem coarse and therefore not virtuous at all!)

Authors who promote this complex attitude, this way of truly being *in* the world, tend to contrast it to a long history of valuing otherworldliness, of seeing Heaven rather than the Earth as our true home. Indeed, the dominant tradition has leaned in this direction; most certainly it has not developed a virtue by the name of earthiness or anything equivalent. Yet the idea of earthiness is not a complete novelty in the long history of Western cultures either. I was struck while reading one Hebrew Bible scholar's description of ancient Israelites as basically earthy people:

> In the Hebrew Bible one encounters a down-to-earth humanness and openness—at times even bluntness and sensuousness—in some of the verbal pictures. Western tastes may be offended at this earthiness. Nevertheless, much of the theology of the Bible comes from terminology rooted in the experiences of everyday life.[69]

By contrast, Christians would less likely be described as earthy people. Yet even in Christian mysticism, that expression of faith which is most intent on attaining communion with God, we can find instances of earthiness. Hildegard of Bingen (1098–1179), for example, devoted much of her energy to studying nature and herbal medicine.[70] Julian of Norwich (1342–c. 1415) discerned God's presence in the experience of a bowel movement.[71] Granted, these are isolated instances, at most a minority tradition. But they necessarily qualify any claim that the appreciation for earthiness would be a completely modern development. Ecological virtues are not always as new as they look, even though some of us are likely to find their innovativeness more interesting than their traditionality.

But then, of course, some of us may actually find their traditional aspects more interesting, or at least more helpful. If an ecological virtue strikes us as familiar in some respect, and if we have not developed an aversion to things traditional, the sense of recognition may motivate us to cultivate this virtue, since we will feel that doing so fits within our understanding of ourselves and our community. Moreover, those of us who are trying to get mainstream constituencies to join the ecological bandwagon may find that presenting ecological virtues in a traditional light reduces the alienation effect. Finally, those of us who are concerned about the deterioration of tradition may see ecological virtues as a viable way to let ancient values speak to modern times.

Although many different motives can lead us to appreciate what is old in what is new, we need to watch for the kind of defensiveness that makes us want to see only what is old. John F. Haught finds such reductionistic defensiveness in what he calls apologetic environmental theology:

> This type of ecological theology . . . argues that if only we practiced the timeless religious virtues we could alleviate the crisis. Since one of the main sources of our predicament is simple human greed, the solution lies in a renewed commitment to humility, to the virtue of detachment, and to the central religious posture of gratitude by which we accept the natural world as God's gift and treat it accordingly. If we allowed our lives to be shaped by genuinely Christian virtues, our relation to nature would have the appropriate balance, and we could avert the disaster that looms before us.

According to Haught, apologetic environmental theology "defends the integrity of biblical religion and traditional theology without requiring their transformation." As such, it does not sufficiently address the current crisis.[72] If Lynn White's insight is correct that we need to *change* Western religious traditions in order to address the ecological crisis at its root, then apologetic reductionism is indeed harmful.

Conclusion

When new worldviews emerge, we can expect significant changes in moral discourse. Ecological virtue language is permeated by the relational and inclusive vision that comes with ecological awareness. Its content as well as its structure reflect this new vision, and as a result the whole discourse has an air of freshness about it.

To those of us, like myself, who used to think that virtues and vices are pretty much timeless, the pervasiveness of this innovation may come as a surprise. When I set out to work on this project, I expected that after scratching the surface a bit, I would find just the old and familiar again. But I did not. No matter how hard I looked, no virtue or vice appeared entirely as it used to, and then there was the amazing profusion of terms I had never encountered in the catalogues of Western traditions. Only the conclusion of a moral paradigm shift could fit my findings. Ecological virtue discourse, I realized, was more like a bird than like a plant: While I had expected to find roots, I found a creature with wings!

As I shifted gears to study fauna rather than flora, I soon realized that my bird was a plant-eater and that I could only get to know it if I also studied what it ate. I did then find some familiar plants, but I now looked at them differently than I had before. I saw how my bird digested them, how they were transformed into a new type of energy, how they were part of a cycle in which everything is old and yet nothing remains the same. Then I understood that ecological virtue language is truly ecological.

NOTES

1. The expression is from Lee H. Yearley, *Mencius and Aquinas: Theories of Virtue and Conceptions of Courage* (Albany, NY: SUNY Press, 1990), 1.
2. I imagine the kind of glue that penetrates and affects the properties of what is glued. This fits the fact that even in a new moral language such as ecological virtue discourse many preexisting terms are used, which only in the process of being put together in a new way begin to take on new meaning.
3. Jay McDaniel uses the term "substantialism." See Jay B. McDaniel, "Christianity and the Need for New Vision," in *Religion and Environmental Crisis*, ed. Eugene C. Hargrove (Athens, GA: University of Georgia Press, 1986), 196.
4. Interesting in this regard is Jim Cheney's charge that the deep ecological idea of self is atomistic, because it resembles one atom rather that being relational ("Ecofeminism and Deep Ecology," *Environmental Ethics* 9 [1987], 130).
5. Douglas C. Bowman, *Beyond the Modern Mind: The Spiritual and Ethical Challenge of the Environmental Crisis* (New York: Pilgrim Press, 1990), 7.
6. Bryan Norton, for example, points to Aldo Leopold's criticism of the atomism of resource management, which "tended to target for management particular species or resources" (Bryan G. Norton, "Should Environmentalists Be Organicists?" *Topoi* 12 [1993], 22).
7. See, for example, Paul Ramsey, *Basic Christian Ethics* (New York: Charles Scribner's Sons, 1950), 355.
8. See, for example, Anne Marie Dalton, "Befriending an Estranged Home," *Religious Education* 85 (Winter 1990), 18.
9. See, for example, Sallie McFague, *Models of God: Theology for an Ecological, Nuclear Age* (Philadelphia, PA: Fortress Press, 1987), 68–9.
10. Like the term "selfishness," the term "anthropocentrism" generally connotes an excessive preoccupation with one's own concerns, at the exclusion of concern for others. This does not mean that the opposite of nonanthropocentrism, like "unselfishness" precludes a healthy care for one's own interests.
11. McFague, *Models of God*, 9. The quote is from Augustine's *Confessions*.
12. See, for example, J. Paul Santmire, *The Travail of Nature: The Ambiguous Ecological Promise of Christian Theology* (Philadelphia, PA: Fortress Press, 1985), 189.
13. McFague, *Models of God*, 4, 11.
14. For the distinction between these two understandings of holism, see J. Baird Callicott, "The Metaphysical Implications of Ecology," in *Nature in Asian Traditions of Thought: Essays in Environmental Philosophy*, eds. J. B. Callicott and R. T. Ames (Albany, NY: SUNY Press, 1989), 61.
15. Thomas Berry, "The Spirituality of the Earth," in *Liberating Life: Contemporary Approaches to Ecological Theology*, eds. C. Birch, W. Eakin, and J. B. McDaniel (Maryknoll, NY: Orbis Books, 1990), 154.
16. Lawrence Johnson, for example, proposes a nonanthropocentric system of species ranking in *Morally Deep World: An Essay on Moral Significance and Environmental Ethics* (Cambridge: Cambridge University Press, 1991). Bryan Norton proposes the following as a central principle of a new postmodern worldview: "All levels of nature are unified by their similar, hierarchical structure and by their common means of maintaining and elaborating that structure through time" ("Should Environmentalists Be Organicists?", 27).
17. Geoffrey R. Lilburne, *A Sense of Place: A Christian Theology of the Land* (Nashville, TN: Abingdon Press, 1989), 49.

18. See, for example, Arne Naess, *Ecology, Community and Lifestyle: Outline of an Ecosophy*, trans. David Rothenberg (Cambridge: Cambridge University Press, 1989), esp. 166ff.
19. Gray, "A Critique of Dominion Theology," in *For Creation's Sake: Preaching, Ecology, and Justice*, ed. Dieter Hessel (Philadelphia: The Geneva Press, 1985), 74, 81–2.
20. Process theologian Charles Birch, for example, argues that one of the meanings of the phrase "the integrity of creation" is: "the interdependence of living organisms and their environment in the complex web of life" ("The Scientific-Environmental Crisis: Where Do the Churches Stand?" *The Ecumenical Review* 40 [April 1988], 191).
21. See, for example, McFague, *Models of God*, 9, 53.
22. See, for example, Ingemar Hedstrom, "Latin America and the Need for a Life-Liberating Theology," in *Liberating Life*, ed. Birch, 112–6.
23. Stanley Hauerwas, *A Community of Character: Toward a Constructive Christian Social Ethic* (Notre Dame, IN: University of Notre Dame Press, 1981), 127.
24. For a parallel critique of narrative ethics based on the work of James M. Gustafson, see William C. French, "Ecological Concern and the Anti-Foundationalist Debates: James Gustafson on Biospheric Constraints," in *The Annual of the Society of Christian Ethics*, ed. Diane M. Yeager (Washington, DC: Georgetown University Press, 1989), 113–30. French argues that Gustafson's critique of anthropocentrism "also challenges the anthropocentrism which is prominant [sic] in many influential accounts of the power of language, texts, stories, and narratives to shape worlds and communities. These do, of course, have a massive impact on the shape of our experience—but so does our genetic heritage and so do the 'orderings' of the biosphere" (114).
25. Walt Whitman, *Leaves of Grass: The 1892 Edition*, with an Introduction by Justin Kaplan (New York: Bantam Books, 1983).
26. This dualistic understanding of virtues and vices can be traced to various formative influences on Christianity, such as Greek (Platonistic) philosophy, Jewish apocalypticism, and Oriental mystery-religions. See, for example, Rosemary Ruether, *New Woman/New Earth: Sexist Ideologies and Human Liberation* (New York: Seabury Press, 1975), 189–90.

    A significant alternative to this dualistic opposition of virtues and vices can be found in the ethics of Aristotle, who saw virtue as a mean between excess and deficiency, and who in addition to virtue and vice recognized four other states of character: godliness, continence, incontinence, and brutishness (*The Nicomachean Ethics*, trans. W. D. Ross [Oxford: Oxford University Press, 1925; reprint ed., 1990], 1145^{a}15–25). Ironically, these nondualistic ideas play a less prominent role in later Thomistic appropriations of Aristotle's work.
27. Waldo Beach and H. Richard Niebuhr, eds., *Christian Ethics. Sources of the Living Tradition* (New York: Ronald Press Company, 1955), 55, 58–61.
28. John Mahoney, *The Making of Moral Theology: A Study of the Roman Catholic Tradition* (Oxford: Clarendon Press, 1987), 7.
29. John Paul II, "Sollicitudo Rei Socialis," *Origins* 17 (3 March 1988), no. 38.
30. Romanus Cessario, O.P., *The Moral Virtues and Theological Ethics* (Notre Dame, IN: University of Notre Dame Press, 1991), 69.
31. See also Erik H. Erikson, *Insight and Responsibility: Lectures on the Ethical Implications of Psychoanalytic Insight* (New York: W. W. Norton, 1964), 111–57.

32. The *results* of vice are, of course, evil insofar as vice harms the biosphere and its parts.
33. For a further discussion on this topic, see chapter 6.
34. The idea that human passions are any more "natural" than other aspects of the human person has been criticized. Stanley Hauerwas, for example, argues that "our passions are as 'unnatural' as our 'reason' and as natural as our reason. Far from being alien to us, and precisely because they are human passions, they make us what we are" (*Community of Character*, 266, note 50).
35. Augustine, *Confessions* (New York: Penguin Classics, 1961), 178.
36. Thomas Aquinas, *Summa Theologiae* I–II.59.5., trans. the Fathers of the English Dominican Province (New York: Benziger Brothers, 1948).
37. Gilbert Meilaender, *The Theory and Practice of Virtue* (Notre Dame, IN: University of Notre Dame Press, 1984), 35, 89, 109, 110. Significantly, Meilaender also describes vices dualistically as "contrary inclinations," though he does not develop the idea (9). He relies here on the work of James D. Wallace (*Virtues and Vices* [Ithaca, NY: Cornell University Press, 1978], 46) and Philippa Foot (*Virtues and Vices and Other Essays in Moral Philosophy* [Berkeley, CA: University of California Press, 1978], 7f.).
38. Jean Porter's recent attempt to build a virtue theory around the idea of "natural perfection" is an important step in this direction. Porter's reconstruction of Thomistic virtue theory parallels the work of those environmentalists who employ a hierarchical understanding of creational continuity. Porter is concerned with showing how Thomistic virtue theory can pay tribute to "the continuity between human beings and the varied world of creatures within which we move." She describes Aquinas's universe as a "web of creatures." (Although the notion of a web is offered as an alternative to hierarchical understandings of creation in much feminist and ecological literature, Porter apparently does not interpret it as such). At the same time, Porter criticizes Aquinas's picture of clearly distinguished species. In line with this view of creational continuity, Porter also highlights Aquinas's idea that human passions, under the direction of reason, form an integral part of virtuous action. Overall, Porter sees reason as perfecting nature. But rather than reading this as an indictment of the imperfection of nature, she reads it as a way to affirm possibilities of natural perfection (*The Recovery of Virtue: The Relevance of Aquinas for Christian Ethics* [Louisville, KY: Westminster/John Knox Press, 1990], 43, 49–50, 54, 60–3, 113).
39. Josef Pieper, *The Four Cardinal Virtues* (Notre Dame, IN: University of Notre Dame Press, 1966), xii.
40. Porter, *The Recovery of Virtue*, 49. See Aquinas, *Summa Theologiae* I.65.2.
41. Porter, *The Recovery of Virtue*, 49. See Aquinas, *Summa Theologiae* I.64.4.
42. Porter, *The Recovery of Virtue*, 62. The reference is to the *Summa Theologiae* of Thomas Aquinas.
43. For example, human waste, not industrial waste, is the greatest source of freshwater pollution. See, for example, Walter H. Corson, ed., *The Global Ecology Handbook* (Boston: Beacon Press, 1990), 155ff.
44. Thomas Berry, "Economics: Its Effect on the Life Systems of the World," in *Thomas Berry and the New Cosmology*, eds. A. Lonergan and C. Richards (Mystic, CT: 23rd Publications, 1987), 9. For a more detailed discussion of Berry's teleological assumptions, see chapter 4.

Whether all of "the natural world" needs to be included in our moral vision, or rather a smaller part, remains a matter of debate. In some literature, the focus is on the entire cosmos. Matthew Fox, for example, calls for "reverence for the cosmos in which we find ourselves" ("Creation-Centered Spirituality from Hildegard of Bingen to Julian of Norwich" in *Cry of the Environment: Rebuilding the Christian Creation Tradition*, eds. Philip N. Joranson and Ken Butigan [Santa Fe, NM: Bear & Company, 1984], 85). Yet the majority of authors are primarily concerned with expanding our ethical horizons simply to the largest system on which human beings have direct and routine impact, which is the biosphere. Interestingly, Eduard Suess, the Austrian biologist who coined the term "biosphere," already described its many relations in moral terms as "the solidarity of all life" (Eduard Suess, *The Face of the Earth* vol. 5, trans. Hertha B. C. Sollas [Oxford: Clarendon Press, 1904–24]).

45. The following examples are drawn from Christopher Key Chapple, ed., *Ecological Prospects: Scientific, Religious, and Aesthetic Perspectives* (Albany, NY: SUNY Press, 1994). The section is an adaptation of my own concluding essay in this volume, "Review and Prospects: The Emergence of a Grounded Virtue Ethic."
46. Rosemary Ruether, "Ecofeminism," in *Ecological Prospects*, ed. Chapple, 165.
47. L. M. Benton, "Selling the Natural or Selling Out? Exploring Environmental Merchandising," *Environmental Ethics* 17 (1995), 21.
48. Albert Fritsch, "Environmental Action Choices," in *Ecological Prospects*, ed. Chapple, 100.
49. As John F. Haught observes, "An ecological spirituality [i.e., a creation-centered, sacramental kind of environmental theology] requires its own kind of *asceticism*. This asceticism prescribes a renunciation not of the natural world but of the Enlightenment ideal of autonomous, isolated selfhood. It subjects us to the arduous discipline of taking into full account the fact of our being inextricably tied into a wider earth-community. A full life, one in which we acknowledge our complex relation to the universe, widens our sense of responsibility toward ourselves and others. Above anything else, this means adopting a continually expanding posture of inclusiveness toward all otherness that we encounter, including the wildness of the natural world" (*The Promise of Nature* [New York: Paulist Press, 1993], 97–8).
50. David Strong suggests that the science of ecology (when using what he calls "disclosive discourse," which engages the reader) has encouraged this kind of respect: "No doubt field ecologists do generate respect for life forms and their connections. What we study long enough has a way of getting under our skin. When sharing their information and stories, they are sharing something of interest" ("Disclosive Discourse, Ecology, and Technology," *Environmental Ethics* 16 [1994], 94).
51. J. Baird Callicott refers to "receptivity of the mind" as a capacity needed for aesthetic appreciation ("The Wilderness Idea Revisited," in *Ecological Prospects*, ed. Chapple, 176).
52. D. Michael Levin, "Transpersonal Experience: Practising Deep Ecology in Education and Self-Formation," paper presented at the 1991 Casassa Conference, Loyola Marymount University, 7, 8.
53. This may explain why it has not been the subject of moral reflection in the past: Changing reality was long seen as less interesting than unchanging reality (God, the Platonic ideals, etc.). Hence it was not particularly important to practice the skills of focusing on what changes.
54. Mary Evelyn Tucker, "An Ecological Cosmology: The Confucian Philosophy of

Material Force," paper presented at the 1991 Casassa Conference, Loyola Marymount University, 4.

The musical example is not just an analogy. I am convinced that the practice of music actually breeds the capacity for attunement, and that the musical origin of the metaphor is no coincidence.

55. For examples of innovation involving vices, see chapter 6.
56. Cf. Hans-Georg Gadamer's discussion of *wirkungsgeschichtliche Bewusstsein* ("effective-historical consciousness," or "historically shaped consciousness") in *Truth and Method* (New York: Crossroad, 1975), 267–74, 305–41.
57. Cf. Yearley, *Mencius and Aquinas*, 1.
58. A good example of such eclecticism can be found in Max Oelschlaeger's book, *Caring for Creation* (New Haven, CT: Yale University Press, 1994). Oelschlaeger claims that the metaphor of "caring for creation" can be supported by various religions, especially by their creation stories (120). Many more examples can be given and it would be interesting also to explore structural parallels between ecological virtue discourse and, say, Confucian virtue ethics or Buddhist virtue ethics. Yet that lies beyond the scope of this book.
59. See the "Statement by Religious Leaders at the Summit on Environment," in *This Sacred Earth: Religion, Nature, Environment*, ed. Roger Gottlieb (New York: Routledge, 1996), 636–9; and "Declaration of the 'Mission to Washington,' " ibid., 640–2.
60. Thus many Western ecowriters have turned their attention to other traditions. Gus diZerega, for example, writes: "How do we respect those life forms, some of whom we must eat? Despite Hume and Darwin, Western ethics gives us little guidance. It is to the deepest teachings of the world's primal peoples that we can most appropriately turn." DiZerega provides several examples of respect in Native American traditions ("Individuality, Human and Natural Communities, and the Foundation of Ethics," *Environmental Ethics* 17 [1995], 33).
61. Interesting in this regard is that the term self-realization explicitly appears in some ecocentric literature. Arne Naess assigns it the following ethical function: "In the systematisation of Ecosophy T [Naess's normative system], the term 'Self-realisation' is used to indicate a kind of perfection. It is conceived as a process but also as an ultimate goal, in a rather special usage of 'ultimate.' It is logically ultimate in a systematic exposition of Ecosophy T. The term includes personal and community self-realisation, but is conceived also to refer to an unfolding of reality as a totality" (Naess, *Ecology, Community, and Lifestyle*, 84). It should be noted that Aristotle did not apply the notion of self-realization to the cosmos as a whole.
62. In addition, Geoffrey B. Frasz and Ron Erickson both explicitly use Aristotle's idea that a virtue represents a mean between two vices as a test to see if the environmental virtues they propose are indeed virtues (Frasz, "Environmental Virtue Ethics: A New Direction for Environmental Ethics," *Environmental Ethics* 15 [1993], 259–74; Ron Erickson, "Comment on Environmental Virtue Ethics," *Environmental Ethics* 16 [1994], 334–6).

As I see it, the use of the doctrine of the mean as a criterion for genuine ecological virtue requires further articulation (see also my comments in chapter 1). For example, it should be clarified whether the mean must be a matter of excess and deficiency of *the same thing* (Erickson's argument does not require this). Also, one would need to account for the existence of virtues that we tend not to perceive as a mean (e.g., love).

63. Aristotle distinguishes between vice and incontinence. Both are based on false knowledge, but "vice is unconscious of itself [since 'the first principle' has been destroyed], incontinence is not" (*Nicomachean Ethics*, $1150^{b}35$). Modern moral discourse, including ecological virtue language, often lumps vice and incontinence together.
64. For Aristotle only "universal judgement" counts as genuine knowledge, even though our experience may go counter to this "right rule" (*Nicomachean Ethics* $1147^{a}25$–$1147^{b}5$).
65. For a further discussion on this point, see chapter 6.
66. See, for example, Aristotle, *Nicomachean Ethics*, $1119^{a}10$–20. Overall, however, when it comes to the role of passions in the life of virtue we see how the structure of ecological virtue discourse is not completely Aristotelian. Environmental authors sometimes suggest that we let our emotions take the lead, that our "body knowledge" may be more trustworthy than our "head knowledge," that we should follow intuition if we are to interact rightly with our environment. This directly counters the views of Aristotle, who in no uncertain terms states that being virtuous means that our intellect must be in charge of our "lower" faculties. Thus despite striking structural similarities, we find a significant difference as well.
67. Not all authors use the term "earthiness" to denote the complex attitude I will describe here. Many use no term at all. Yet I find it useful to put a label on a combination of traits that I do think can be categorized—and widely recognized—as a single virtue.
68. Fritsch, "Environmental Action Choices," in *Ecological Prospects*, ed. Chapple, 94.
69. Marvin R. Wilson, *Our Father Abraham: Jewish Roots of the Christian Faith* (Grand Rapids, MI: Eerdmans, 1989), 146–7.
70. See Gabriele Uhlein, *Meditations with Hildegard of Bingen* (Santa Fe, NM: Bear & Co., 1982); Wighard Strehlow and Gottfried Hertzka, *Hildegard of Bingen's Medicine*, trans. Karin Anderson Strehlow (Santa Fe, NM: Bear & Co., 1988). On earthiness in medieval mysticism, see Matthew Fox, "Creation-Centered Spirituality," in *Cry of the Environment*, ed. Joranson, 85–106.
71. Julian of Norwich, *Showings*, trans. Edmund Colledge, O.S.A. and James Walsh, S.J. (New York: Paulist Press, 1978), 186.
72. Haught, *The Promise of Nature*, 91–3.

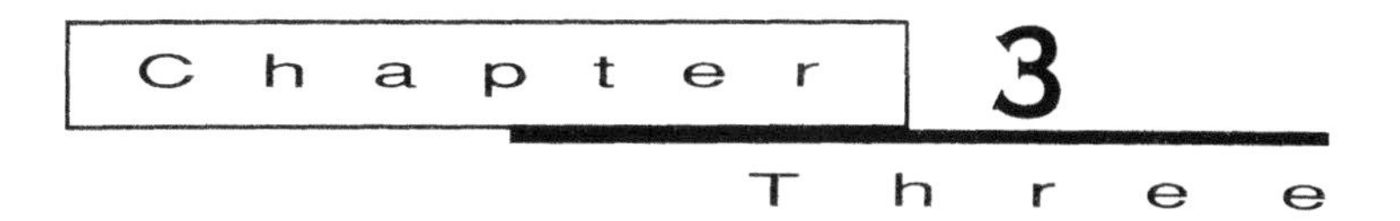

THE VIRTUE ETHIC OF MURRAY BOOKCHIN

IN THE PREVIOUS CHAPTERS we took a rather fast tour through the territory of ecological virtue language. I have pointed out general features of the discourse, emphasizing unique as well as more familiar aspects. Now it is time to settle down and appreciate some details. Since we cannot see everything close-up, I will focus on the virtue discourse of two remarkable (and remarkably different) authors: Murray Bookchin and Thomas Berry. Both are seasoned and experienced writers. Their works carry the mark of decades of reflection and active commitment to shaping a better world. Rich virtue language permeates the writings of each. Yet the decades of reflection and action have taken Bookchin and Berry in rather distinctive directions. (What else could we expect of creative thinkers?) Their quite explicit worldviews differ significantly, and these differences are reflected in their virtue languages as well. Thus, between this chapter on Bookchin and the next chapter on Berry, we will gain a detailed understanding of certain tensions within the emerging ecological virtue discourse, tensions that raise questions for further reflection.

Murray Bookchin has long been known as a spearhead of ecological consciousness in the United States. Already in the 1950s, Bookchin alerted his readers to environmental problems and developed an ecological perspective based on social criticism.[1] Bookchin's comprehensive theory, known by the name of social ecology, represents a distinct approach within environmental philosophy.[2] It combines libertarian anarchism with a dialectical understanding of nature that is presented as an alternative to both biocentrism and anthropocentrism. Bookchin's work has recently been the focus of much debate, occasioned by his criticism of what he sees as misanthropic tendencies in the deep ecology movement.[3] While the debate has highlighted the important question of the role of humans in relation to nonhuman nature, it has overshadowed a dimension of social ecology that deserves more attention, namely, the rich and in many ways uncommon virtue ethic that permeates the work of Murray Bookchin. In this chapter I will provide an analysis of the virtue ethic of *The Ecology of Freedom*, the work Bookchin considers the "keystone" of his "theoretical arch."[4]

In *The Ecology of Freedom*, Murray Bookchin presents the thesis that human domination of nonhuman nature is rooted in human domination of other humans, a thesis that inverts the common view of the socialist tradition to which Bookchin is heir. In addition, the work develops a utopian vision of "a reharmonization of nature and humanity through a reharmonization of human with human" (11). Thus, *The Ecology of Freedom* is about a moral project.

The language that gives expression to this moral project is the language of virtue. Although Bookchin does not explicitly identify his ethic as a virtue ethic, he does indicate a preference for virtues over rights. Moreover, while he criticizes the utilitarian theories of Bentham and Mill, he expresses respect for the "classical ideal" of virtue as well as for certain aspects of the American "republican virtue" tradition.[5] In *Remaking Society: Pathways to a Green Future*, a later work that summarizes many of the insights from *The Ecology of Freedom*, Bookchin characterizes his own approach as "an ethics of the greater good and virtue," contrasted to "an ethics of lesser evils," which he finds in reformist environmentalism.[6]

Although *The Ecology of Freedom* does not contain an explicit virtue theory (the focus of the book lies elsewhere), elements of such a theory can be gleaned from the text. In structure, Bookchin's ethic is Aristotelian. The life of virtue derives direction from a clearly articulated teleological concept of the human good.[7] Reason plays an important role (though not the only role) in the achievement of this good[8]: It helps us both articulate the good and discern the good in practical situations. The latter is often a matter of finding the middle course between harmful extremes.[9]

However, it is not the conventional Aristotelian features of Bookchin's virtue ethic that make it interesting, but rather its unconventional elements. When it comes to the question of adapting Western virtue traditions to the insights and demands of a postcommunist, ecological age, one can find much food for thought in *The Ecology of Freedom*. I will demonstrate this by discussing Murray Bookchin's contributions in shaping an ecological worldview that impacts the life of virtue, in articulating an innovative catalogue of virtues and vices, and in explaining how the cultivation of virtue relates to social change.

Vision and Virtue in Social Ecology

As I mentioned in chapter 1, many ecowriters have criticized the hierarchical dualisms they find in Western traditions. The work of Murray Bookchin provides a particularly well-developed example. Bookchin also offers a clear alternative worldview, based on the notion of dynamic differentiation within continuity.

From the perspective of virtue theory, it is interesting to find such an ecological worldview in conjunction with a rich virtue ethic. One wonders what difference having a nondualistic, nonhierarchical, dynamic worldview could make for the life of virtue, which has so often been conceived in terms of static hierarchical dualisms.[10] In this section I will spell out some of these implications. Also interesting from the perspective of virtue ethics is the fact that Bookchin's worldview remains in many ways Aristotelian, especially in its teleological orientation. Thus his work provides a good overall test case for the adaptation of Aristotelian virtue ethics to an ecological age.

In *The Ecology of Freedom*, we primarily find a critique of society/nature and mind/matter dualisms (e.g., 38, 279). Bookchin's critique of these dualisms is an outgrowth of his social theory; it does not reflect the desire for mystical unification that marks the critiques of many other authors. Bookchin maintains an interest in making clear distinctions, especially where social and moral matters are concerned. From his work we learn that a holistic worldview need not imply lack of moral discrimination, and that making moral distinctions does not, by definition, commit us to a dualistic worldview:

> That there are phenomena in our world that must be conquered, indeed, disgorged—for example, domination, exploitation, rule, cruelty, and indifference to suffering—needs hardly to be emphasized. But that "otherness" *per se* is intrinsically comprehended in oppositional terms also biases that comprehension in the direction of instrumentalism, for hidden within a dialectic of strict negativity are the philosophical tricks for using power as a predominant mode of comprehension. (302)

Between the extremes of symbiotic unification and instrumentalist dualism, Bookchin locates the desirable option of seeing the world as a variegated network. Within this network, sensible and necessary distinctions can be made. For example, it still makes sense to distinguish between virtues and vices, as Bookchin often does.

The Ecology of Freedom presents dualistic thinking as an ideology in the service of domination. Consequently, we find a critique of hierarchy underlying the critique of dualism. Bookchin defines hierarchy as "the cultural, traditional and psychological systems of obedience and command, not merely the economic and political systems to which the terms class and State most appropriately refer" (4).[11] In addition to being a social phenomenon, hierarchy is a state of consciousness (often referred to as "hierarchical sensibility"; the term is used in the role of a capital vice[12]). Bookchin stresses emphatically that hierarchy is not found in nature but must be seen as a strictly social phenomenon with roots going back to tensions inherent in primordial hunting and gathering communities (6). The tendency of our traditions to perceive a hierarchical order in the natural world is thus an instance of projection. This projection is not as innocent as it may seem, for in addition to denying "the integrity of nature" it frequently serves to justify social domination. Thus the "king of the beasts" is not a benign image (Disney's *Lion King* notwithstanding). Nor is the pyramidical representation of the food chain, which has, in turn, served to justify human domination of the natural world (26–8).

The critique of hierarchy stands at the core of *The Ecology of Freedom*. It has far-reaching implications for virtue ethics. As I already mentioned in chapter 2, critiques of hierarchy imply that the fearful link between vice and animal behavior must be questioned. Bookchin's specific critique presents an additional a priori challenge to the validity of virtue traditions that have been developed or elaborated to serve elite interests (and there are several).[13] Moreover, if Bookchin's premises are on target, they would undercut those traditions that have grafted a virtue ethic on a hierarchical understanding of the order of nature, projected back onto human nature. The old "ladder" of virtues would have to be revised, the association between virtue and nobility rethought, the role of reason in the cultivation of virtue recast, and the life of virtue liberated from the repressive dynamics of obedience and guilt.

As an alternative to dualistic and hierarchical thinking, Bookchin offers a holistic worldview of the "weak" kind, based on the notion of differentiation within continuity.[14] There is continuity between nature and society, since one has gradually emerged from the other and remains rooted in it (279). There is also continuity within nature, since all of its parts—including us humans—are connected through food webs (26). And there is graduated continuity between matter and subjectivity (11, 235). For Bookchin, holism implies that everything has a common origin and is intricately connected in a functional network; it does not refer to a single spirit or essence permeating reality. His holism is also dynamic: The world develops in the direction of ever greater differentiation, but since the differentiated parts have a single origin and develop in relation to each other, there always remains a degree of connection.

For Bookchin this differentiated human/nonhuman continuity is a good in the moral sense, and it develops in a morally desirable direction. Here the teleological character of Bookchin's theory comes into view. Bookchin, in fact, acknowledges his indebtedness to the Aristotelian notion of *entelechial* causation (283). He warns, however, that the term "teleological" must be used with caution, for it "is redolent with notions of a predetermined, inexorable end—a notion that Aristotle takes great pains to eschew" (284). Bookchin is tentative, suggesting only that when we observe the evolutionary development of the natural world, we discern a certain directiveness that has consistently pointed toward ever greater variety and subjectivity. This directiveness is a realization of potentialities that have been present from the very beginning, a process Bookchin finds hard to attribute to "mere fortuity." Thus, he admits, "we cannot help but speculate about the existence of a broadly conceived *telos* and a latent subjectivity in substance itself that eventually yields mind and intellectuality" (364).[15] The assumption that there is such a telos forms the basis of Bookchin's ethics. It supports the idea that nature can be a source of moral meaning, as well as his utopian view of human destiny. Moreover, as we shall see in chapter 5, it provides an ultimate standard for determining what is genuinely virtuous.

The overall telos of natural development includes a human telos that is distinguishable from the goals of natural ecosystems.[16] Human destiny belongs to, and presupposes, the evolutionary self-realization of nature, but within this larger process humans have a unique place. A natural ecosystem (the development of which can be seen as a snapshot in a larger evolutionary movement) "finds its climax in the greatest degree of stability it can attain within its given level of possibilities," a stability attained through maximum diversification, as in a climax forest (36). Thus the climax of a concrete ecosystem represents a particular high point in the evolutionary movement toward ever greater variety.[17] But the potential for subjectivity remains underdeveloped in a climax ecosystem of which humans are not a part. This potential can, however, find full expression in a human ecocommunity. Resting on the stability of its ecological context, the human social realm can reach a high point in the development of subjectivity when it achieves *freedom and self-consciousness* (36). Thus, by developing their unique potentialities for consciousness, humans not only reach a specific end that *contributes* to the realization of a larger end (variety),[18] but they also reach a specific end that *coincides with* the realization of a larger end (subjectivity).[19] This is crucial, for it means that humans find an ultimate end in themselves; they are not just functional pegs that serve a system (though they are that too). By making these distinctions, Bookchin avoids falling into the fascist trap that has haunted much teleological thinking (of which he is all too aware).

If hierarchy is the focus of Bookchin's critique, freedom is the focus of his utopianism. This human telos has various dimensions. It implies "the full realization of humanity's potentialities in their most creative form" (10), especially "the achievement of an autonomous personality and selfhood" (70) and the ability "*to shape material life in a form that is . . . ecological, national, and artistic*" (218). Socially, it expresses itself as "the equality of unequals," a society in which all receive according to their unique needs (9). Thus freedom is "more than the absence of constraint" (37), though that is certainly presupposed. When not repressed or constrained, humans instinctively move toward this telos (195).

If a virtue is a disposition that makes both humans and their actions good, *The Ecology of Freedom* gives concrete content to this definition.[20] The good is human freedom and self-consciousness embedded in a larger context of natural stability and diversification. Thus Bookchin provides an ultimate and inclusive standard—in fact, a unity of differentiated teloi—from which the life of virtue could (if his conceptual framework stands up in further philosophical scrutiny) take its bearings in an ecological age.

At the same time, Bookchin's proposed standard leaves room for variation. Creativity and personality, both expressions of freedom, by definition imply open-endedness and uniqueness. This is where we reach the limits of the general vision Bookchin offers for the life of virtue, a point at which actual historical and personal experiences must begin to play a more explicit role. It is only in the interplay between a sense of telos and a sense of living tradition that a concrete virtue ethic can be articulated. I will now turn to the concrete virtues and vices of *The Ecology of Freedom*.

A Distinctive Catalogue

At various points in *The Ecology of Freedom*, Murray Bookchin presents his readers with short lists of virtues and vices. For example, "cooperation, mutual support, and love" (317) are mentioned in one breath, as are "domination, self-interest, and exploitation" (307). Yet Bookchin does not integrate the various lists, nor do they by themselves reflect the full variety of virtue language that can be found in *The Ecology of Freedom*. In order to show this richness and at the same time to allow comparison with the general overview of ecological virtue discourse provided in Appendix A, I have compiled a catalogue that represents Bookchin's virtue ethic.

The following concepts (or derivative terms, depending on the grammatical context) occur in the role of virtues in *The Ecology of Freedom*[21] (concepts most frequently mentioned and/or particularly stressed by the author are printed in italic type):

attunement
autonomy/disengagement
care
carefulness/cautiousness
complementarity
conciliatory sensibility
cooperation
courage/heroism
creativity
earthiness
ecological sensibility/libertarian rationality
generosity
gentleness
grace/elegance/delicacy
healing
holistic thinking
honor/sense of moral probity

hope
individuality/having personality
intellectual sophistication
interdependence
kindness
love
managing
maturity/emotional fortitude
nurturing
organic knowledge
participation
prudence
rebelliousness/liberatory sensibility
respect
responsibility
self-reflexivity
sensitivity
sensuousness/passion for the concrete
sharing
solidarity/sense of unity
spontaneity
supporting
sympathy
temperance/restraint
theoretical discrimination
vision/utopian thinking
wisdom

The following concepts occur in the role of vices in *The Ecology of Freedom*:[22]

aggressiveness
alienation
arrogance
atavism
authoritarianism/authoritarian rationality
blindness
coerciveness
competitiveness
consumerism
cruelty
denial
ecofascism
egotism/self-interest
elitism
emotional fragility/acting spoiled
environmentalism
envy
exploitation
fanaticism
fatalism
forcing
greed/avarice
hatred
hierarchical sensibility
hubris
ignorance
instrumental rationality/ instrumentalism
irrationality
lack of imagination
manipulation
mastery/*domination/oppression*
passivity
pessimism
reductionism
repression
self-deception
sentimentality
stupidity
wishful thinking

This catalogue shows a relative balance between virtues (44 in total) and vices (39), even though there is no attempt by Bookchin to match them systematically. In the text, however, the virtues tend to receive much more discussion than the vices. For all of his strident style and biting condemnations (which have at times drawn more attention from critics than the substance of his thought), Bookchin actually presents us with an ethic that on the whole is positive and constructive.

The catalogue contains some very traditional terms (love, hope; hatred, greed), combined with terms that reflect a more modern ethic (creativity, solidarity; domination, repression).[23] Certain terms reflect the specific concerns of many in the environmental movement (respect [for nature], exploitation [of nature]). Others bear the mark of Bookchin's unique approach to social and ecological issues (utopian thinking; hierarchical sensibility). The ostensibly eclectic result is not arbitrary, however. Bookchin is very aware that Western virtue tradition, including the Aristotelian tradition, is a mixed bag that continues to change.[24] His selection of virtues and vices reflects a careful and reasoned attempt to salvage the best of this tradition and its recent innovations while supplementing it where necessary. In chapter 5 I shall draw out four criteria that mark Bookchin's selection and show how these can be used as a springboard for extending Aristotelian virtue tradition to meet the demands of an ecological age.

A catalogue, of course, provides only the bare bones of a virtue ethic. In the remainder of this section I will put some meat on these bare bones by highlighting two features that characterize the way in which Murray Bookchin understands many of the virtues and vices that occur in *The Ecology of Freedom*, namely: (1) the exemplary role of tribal life; and (2) the ambivalent role of socialist themes.

For Bookchin, prehistoric as well as modern tribal life functions as a concrete model to which he frequently refers when he wants to illustrate what he has in mind in describing the virtues that belong to an ecological society. In order to show what he means by care, for example, Bookchin points his readers to the communal solidarity and mutual care of New World prehistoric societies, as well as to the Eskimos, who "crafted their equipment with considerable care because they had a high sense of care for each other" (242). Artistic objects from the past attest to a "creative subjectivity" and "attuned sensitivity" that modern people must learn to regain (241). The meaning of true mutual respect is illustrated by the language and practices of the Wintu Indians, who use cooperative terms for expressions that in our language denote coercion (e.g., "to rule" is "to stand with"; 45). In general, Bookchin points out, one finds much respect between the sexes in primal communities (53). Organic societies also show "a community solidarity that such widely bandied terms as stewardship can hardly convey" (48). Such organic solidarity also extends to the natural world (44, 48). These examples show that Bookchin has quite a concrete image in mind when he describes the virtues he finds important. One could say that his virtue ethic, though not conceptually developed, is fleshed out by means of historical examples.

Still, Bookchin does not uncritically accept all that tribal life has brought forth. This is particularly clear from his descriptions of vices. For example,

Bookchin denounces the harshly aggressive outlook of patriarchy, which also latently characterizes certain tribal cultures (6), and he deplores the fact that in some societies "masculine" aggression is given mystical sustenance by shamans, who often end up manipulating people (85, 100). Bookchin traces hatred of nature, of *eros*, and of the body to the fear of the aged, who in primordial societies were especially susceptible to "nature's vicissitudes" and who at the same time were able to acquire a position of power (gerontocracy) that enabled their ambiguous attitude toward nature to be institutionalized (82). In general one can conclude that tribal societies that have become *hierarchical* do not fulfill a model function in the virtue ethic of *The Ecology of Freedom*. Bookchin's anarchism, to which his affinity for tribal societies may well be ascribed, provides a critical perspective that ensures that not every feature of tribal life will be held up as a model for the life of virtue.

This brings me to my second point, namely, the role of socialist themes in the interpretation of particular virtues and vices. Bookchin describes himself as a libertarian anarchist, and as such he is clearly ambivalent toward the larger socialist legacy of thought. This ambivalence also marks his interpretation of virtues and vices. On the one hand, the virtue ethic of *The Ecology of Freedom* at certain points reflects typically socialist concerns. For example, Bookchin has a keen interest in the nature of labor, and because of this interest he often links the virtue of creativity to the process of production (e.g., 227, 266). Moreover, the key socialist concept of solidarity is a central virtue in Bookchin's ethic. And Bookchin's critique of social domination as well as organized religion, which also expresses the intellectual heritage of socialism, leads him to treat humility as a counterfeit virtue (whereas this is a key virtue for many in the environmental movement).

On the other hand, Bookchin also explicitly rejects some key socialist ideas, and this critical attitude is reflected in his virtue ethic as well. For example, he rejects Marx's idea that social domination results from an inevitable human need to dominate nature. Bookchin turns the causality around: Attempts to dominate external nature occur "when humanity is conceptually equipped to tranfer its social antagonisms to the natural world outside" (82).[25] Because of this reversal, the domination of nature becomes a vice (which it clearly is not for Marx).

Overall, at the nuts-and-bolts level of concrete virtues and vices, *The Ecology of Freedom* provides a richly varied ethic with its own distinctive character. As an anarchist, Bookchin is inspired by both tribal life and the legacy of socialism. Yet his anarchism also ensures that he maintains a critical perspective toward both. This combination of features influences his ethic in a very concrete way and makes it relevant to modern issues of social justice as well as ecological sustainability.

Virtue and Social Change

Not only does *The Ecology of Freedom* contain a moral vision and a concrete set of virtues and vices for an ecological age, but it also puts the cultivation of virtue in the larger perspective of the question of social change. Bookchin's contribution in this area stands out among other environmental literature due to his careful study of the dynamics of social change and the role of moral values in this process. Bookchin does not follow the simple assumption, outlined in chapter 1, that the main avenue to social change is a transformation of personal attitudes. Yet he does not deny it either. Rather, he places the transformation of attitudes into a wider theory of social change that reflects an unusual cross-fertilization between Marxist dialectics and Aristotelian political philosophy.

How we get "from here to there," from a destructive social system to one that supports human and nonhuman self-realization, is a question that is always on Murray Bookchin's mind. Finding an answer to this question requires first of all social criticism, a thorough understanding of what went wrong, and why.[26] Again, Bookchin's answers follow a middle course: Current social and ecological problems can neither be entirely blamed on human vices, nor can they be entirely explained in economic terms. Thus Bookchin steers between the Scylla and Charybdis of traditional Christian and Marxist reductionism.[27]

This middle course is particularly interesting in light of the discussion that has rocked the environmental movement since the publication of Lynn White's essay, "The Historical Roots of Our Ecologic Crisis." White blamed current ecological problems on an "attitude" of domination fostered by Christian tradition. While the ensuing debate has largely focused on whether the Christian tradition indeed carries this "burden of guilt," the underlying assumption that an "attitude" is the main cause behind the crisis has been rather unquestioningly accepted by many ecowriters. Bookchin's position that no current social problem is simply attributable to human vice thus stands out. In fact, in *Remaking Society* Bookchin explicitly criticizes White's social analysis, leaving no doubt about his own views.

> More than ever, strong emphasis must be placed on the fact that nearly *all ecological problems are social problems*, not simply or primarily the result of religious, spiritual, or political ideologies. That these ideologies may foster an anti-ecological outlook in people of all strata hardly requires emphasis. But rather than simply take ideologies at their face value, it is crucial for us to ask from whence these ideologies develop.[28]

This passage is particularly revealing. Bookchin, for whom domination is clearly the most capital of vices, is nevertheless not willing to say that the (alleged) Christian promotion of domination lies at the root of the current ecological crisis. His Marxist consciousness prevents him from drawing such a conclusion. Vices may be pernicious, but *at this stage in human history* they are not the only corrupting influence. Entrenched social and economic structures (though themselves to a degree sedimented vices) play a more important role.

At the same time, however, it would be a mistake altogether to neglect vices in social criticism. Bookchin sees attempts to explain everything in terms of economic forces as an outgrowth of capitalism (in this respect even Marx was a child of his time):[29]

> Almost every critique of the "bourgeois traits" of modern society, technics, and individuality is itself tainted by the very substance it criticizes. By emphasizing economics, class interest, and the "material substrate" of society as such, such critiques are the bearers of the very "bourgeois traits" they purport to oppose. They are in perilous default of their commitment to transcend the economic conditions of capitalist society and to recover the ethical level of discourse and ideals capitalism so savagely degraded. (217)

To get stuck on Marx in our social criticism—which, incidentally, happens less nowadays than when *The Ecology of Freedom* was written—is ultimately as myopic as to be ignorant of him. Bookchin proposes that we move *through* Marx and take into account the active role of human vice in history.[30] Only then can we engage in balanced social criticism that gives due recognition to economic factors without slighting human moral agency and responsibility by reducing everything to a material base. Only then can we reasonably hope to be effective in avoiding the mistakes of the past.

A good example of Bookchin's balanced social criticism can be found in his discussion of the historical role of psychological repression. Bookchin frequently refers to the dynamics of repression in *The Ecology of Freedom*. Repression functions as a capital vice, giving rise to many other vices and evils. One of these is the exploitation of nonhuman nature:

> The discipline of *rule* . . . demands the repression of internal nature. This repression then extends outward to external nature as a mere object of rule and later of exploitation. Unless we explore this history, which lives actively within us like earlier phases of our individual lives, we will never be free of its hold. We may eliminate classes and exploitation, but we will not be spared from the trammels of hierarchy and domination. (8)

This passage nicely illustrates the interaction between repression and socio-economic life. The vice of repression, itself a result of oppressive social structures, leads to the oppression and exploitation of nature by society. We cannot get rid of the final problem without paying attention to the dynamics of vice that contributed to it.

Just as current social and ecological problems are not caused by human vices or economic forces alone, so social change cannot be achieved simply by changing attitudes or simply through technological and economic innovation. On the one hand, to call for new ecological virtues and lifestyles while ignoring the need to reform corporations and social structures is to look for piecemeal "solutions" that are essentially reactionary.[31] With a critical eye to deep ecology, Bookchin warns: "What is of *pivotal* importance in determining whether human beings will creatively foster the evolution of first nature or whether they will be highly destructive to nonhuman and human beings alike is precisely the kind of *society* we establish, not only the kind of sensibility we develop" (Introduction to 1991 edition, xxxii). We must be careful not to "retreat to strategies of personal 'self-transformation' and 'enrichment,'" as is increasingly the case in radical environmental circles (il). Although there is room for these strategies, they must not become substitutes for social action.

On the other hand, achieving social change is also not merely a matter of changing social laws and institutions, technology, and economic structures. Bookchin believes we can learn from history on this score. Looking back to the Middle Ages, he observes:

> To attain a "Christian" society . . . not only did systems of ownership and the distribution of goods have to be changed, but even as late as Reformation times, "matters of the soul"—the accepted mores, beliefs, institutions, and in a more personal vein, one's character and sexual life—required alteration. These broader needs—indeed this view of need itself—cannot be reduced to mere "superstructural" ideologies without forcing the mentality of a market society on a largely manorial one. (216)[32]

Today, Bookchin finds the same reductionism in reformist environmentalism. It is inadequate in his eyes because it will fail to achieve anything that resembles Aristotle's notion of the "good life," which goes beyond material security and includes ethical ideals of commitment to the well-being of family, friends, and polis (221).

Even as the cultivation of virtue is necessary to bring about social change, so are certain social and technological conditions necessary as *prerequisites* for the cultivation of virtue. Bookchin asks, "What institutions will be required to create a new public sphere, what social relations to foster a new ecological sensibility, what forms of work to render human practice playful and creative?" (40). *The Ecology of Freedom* contains many practical suggestions.[33] For example, in order to have articulate moral agents who can freely pursue genuine personal, social, and ecological goals, we need a decentralized political system based on direct democracy (336–8).[34] Moreover, in order to have individuality and personality, we need mother-infant and family relationships marked by care, cooperation, and love (317). And in order to have sensitivity to nature and a nonhierarchical sensibility, we need an ecological technology based on organic systems (265, 267, 346). All of these are social prerequisites for the life of virtue.

But if we need virtues to build an ecological society while also needing an ecological society to cultivate genuine virtues, does this not mean that we find ourselves in a chicken-egg situation? The logic of *The Ecology of Freedom* suggests that this is not the case. Imperfect, historically grown social structures do actually provide a beginning point for the process of change in the form of small seedbeds of genuine virtue. What is needed is a careful redemption of these seedbeds from the surrounding polluted soil of harmful social structures:

> We need to surmount the evil that lies in every good, to redeem the gain that inheres in every loss—be it the sociality latent in the solidarity of kinship, the rationality in primal innocence, the ideals in social conflict, the willfulness in patriarchy, the personality in individualism, the sense of humanity in the parochial tribal community, the ecological sensibility in nature idolatry, or the technics in shamanistic manipulation. To redeem these desiderata without completely shredding certain features of the context that gave them viability—solidarity, innocence, tradition, community, and nature—will require all the wisdom and artfulness we possess. Nor can they be adequately redeemed within the present social order. (268–9)

As a whole, the current social system cannot support the cultivation of ecological virtues. Yet parts of this system do provide viable and necessary conditions for the life of virtue, if only we knew how to recognize them and put them together in a new way. For example, even within the environmentally and socially destructive structures of suburban sprawl, back gardens, neighborhood meetings, and parent cooperatives provide real conditions for an alternative life of virtue. Our task is to seek out these moral seedbeds and to dwell there, so that we ourselves can grow and the seedbeds flourish and expand. Thus the new virtuous person and the new social structure will emerge simultaneously. This revolution will happen slowly, but with a strength and attraction before which the old will ultimately crumble.[35]

CONCLUSION

At the end of the street where I grew up there was an empty lot. One day a truckload of street bricks appeared (roads are often paved in the Netherlands). To us kids it seemed as though they had fallen straight from heaven. We began to build our dreamworld—that is, the guys did, as well as my best friend, a tomboy. The guys assured me they would build the most beautiful house of all for me, for I was "a lady" and deserved special treatment. I could not see the house until it was ready. When it was, I only felt disappointment (though I did not show it). The spacious brick hut with flowers tucked in the sandy corners had nothing to do with *me*! I must have been as ill at ease as the neighborhood dog, for whom they had built a special cage. A few days later we came back from school to find the village gone. The powers that be had leveled it and paved a parking lot.

Reading *The Ecology of Freedom*, I was reminded of this sweet-and-sour experience. We had built the external structure of a little utopia, but we had not renewed our relationships and sensibilities, nor had we paid attention to the plans and interests of "the big world outside." So our dreamworld was not to be a utopia for all, and in the end even the external structure was destroyed. In an age marked by serious social and ecological problems, we cannot afford such naiveté. We need to renew the life of virtue as well as work for the renewal of social institutions, and preferably we should integrate these efforts. At the academic level this means that a greater integration between virtue theory and social theory and ethics is needed.

The Ecology of Freedom provides many specific ideas for such an integration. It contains a rich and concrete virtue ethic built on a vision that reflects ecological insights as well as an awareness of society's potential for harm. This ethic is rooted in tradition and yet at many points carefully distinguished from it, yielding a concrete model that could help us in the critical task of discerning where to go from here. Finally, Bookchin's ethic is integrated with a theory of social change that in turn addresses the societal, economic, and technological conditions for the cultivation of virtue.

"Rarely is history notable for its capacity to select and preserve the most virtuous traits of humanity" (51). Murray Bookchin is a realist when it comes to the actual shape the life of virtue has taken in the past. Yet he is not fatalistic:

> There is still no reason why hope, reinforced by consciousness and redolent with ancestral memories, may not linger within us as an awareness of what humanity has been in the past and what it can become in the future. (51–2)

Bookchin is defiantly hopeful about the life of virtue, showing above all that even in a worsening ecological situation it makes sense to try, and that our efforts can have a more solid basis than random trial and error.

NOTES

1. See Murray Bookchin [Lewis Herber], "The Problem of Chemicals in Food," *Contemporary Issues* 3, no. 12 (1952), 206–41.
2. Bookchin is the best-known social ecologist. Some other authors are Rudolf Bahro, Janet Biehl, John Clark, Andre Gorz, and Chaia Heller. Social ecology is especially developed at the Institute for Social Ecology in Plainfield, Vermont. The institute publishes the journal *Harbinger: The Journal of Social Ecology*.
3. See especially Steve Chase, ed., *Defending the Earth: A Dialogue Between Murray Bookchin & Dave Foreman* (Boston: South End Press, 1991).
4. Murray Bookchin, *The Ecology of Freedom: The Emergence and Dissolution of Hierarchy* (Palo Alto, CA: Cheshire Books, 1982). References to *The Ecology of Freedom* will appear in parentheses in the text of this chapter; in the notes, I will use the abbreviation *EF*. In 1991 a new edition was published (Montreal: Black Rose Books). In the introduction to this edition Bookchin reaffirms most of the ideas in the book and indicates the central place of *The Ecology of Freedom* in his work (xvi).
5. Ibid., 295. On Jeremy Bentham, see 163–5; on John Stuart Mill, see 165.
6. Murray Bookchin, *Remaking Society: Pathways to a Green Future* (Boston: South End Press, 1990), 14. In the new introduction to *The Ecology of Freedom* he also describes his critical historical inquiry in terms of "unscrambling the virtues and vices of humanity's social development" (xlviii).
7. This teleological development does not extend to a "supernatural" goal, as it does in Thomistic virtue ethics. Bookchin stands in the part of Western virtue tradition that ties the life of virtue to a this-worldly good. He does not hide his disdain for "the class and priestly emphasis on the supernatural, with its after-life rewards for obedience and virtue" (*EF* 175).
8. In this regard, Bookchin also sees himself as standing in a long tradition of social theory: "Every appeal of human consciousness, be it 'class consciousness' or 'personal consciousness,' is an appeal to the creativity of mind and an expression of belief in human virtue. Marx the 'materialist,' Hegel 'the idealist,' Kropotkin the 'ecologist,' and Fourier the 'utopian' have all embarked on the same voyage of hope: a belief in the powers of human reason to attain a free society" (*EF* 217–8).
9. Bookchin acknowledges his indebtedness to Aristotle on this point: "If happiness was a rational and virtuous way of life, as Aristotle argued, it attained its full realization in the contemplative mind and in an ethical mean that rose above excess of any kind" (*EF* 165). For Bookchin, a (secular) Jew, reason's function of determining the mean takes on particular urgency in the context of Nazi excesses committed during the Second World War. It explains his profound uneasiness with what he perceives as a tendency by some environmentalists to correct the mistakes of the past by "swinging mindlessly from one extreme that advocates the complete 'domination of nature' by 'man' to another, rather confused 'biocentric' or antihumanist extreme" (*Remaking Society*, 13).

   Reason itself too is subject to a mean: "To return to irrationality, superstition, and material primitivism [in our efforts to deal with current social and ecological problems] is no more desirable than to defer to the value-free and elitist rationalism, scientism, and technocratic sensibilities that prevail today" (*EF* 268). Elsewhere Bookchin subscribes to the rational ideal of "theoretical discrimination," which lies between the extremes of reductionism and dualistic thinking (*EF* 31).

10. For a discussion related to this question, see John H. Riker, *Human Excellence and an Ecological Conception of the Psyche* (Albany, NY: SUNY Press, 1991).
11. Hierarchy does not always imply the existence of exploitation. In *Remaking Society*, Bookchin distinguishes hierarchy from "personal, individual, and often short-lived differences in status that may, in all too many cases, involve no acts of compulsion" (29).
12. I use the term "capital vice" in the sense in which Thomas Aquinas uses it, namely, as a vice that gives rise to many other vices (see *Summa Theologiae* I–II.84.3: "A capital vice is one from which other vices arise, chiefly by being their final cause"). This definition, in contrast to common perception, is not hierarchical but rather developmental: A capital vice is not the "basest" of sins, but rather a particularly pernicious sin. Thus the concept of a capital vice is consistent with Bookchin's nonhierarchical, developmental theory.
13. A similar challenge has been raised on the basis of feminist critiques of patriarchy. See, for example, Susan Moller Okin, *Women in Western Political Thought* (Princeton: Princeton University Press, 1979), esp. chap. 4; Mary Daly, *Pure Lust: Elemental Feminist Philosophy* (San Francisco: Harper, 1984), 218ff.
14. Bookchin has named this view "dialectical naturalism." It is most extensively discussed in Murray Bookchin, *The Philosophy of Social Ecology: Essays on Dialectical Naturalism* (Montreal: Black Rose Books, 1990).
15. In articulating this teleological vision, Bookchin is inspired by Hans Jonas, *The Phenomenon of Life* (New York: Delta Books, 1966).

    There is a certain ambiguity in *The Ecology of Freedom* as to the precise content of this telos. Whereas the telos is often exclusively defined in terms of subjectivity, there are other passages that suggest that variety, which in turn serves stability, is also part of the telos. In that case the telos would be *multidimensional*. In this chapter I will follow the latter option, which is both more nuanced and less vulnerable to promoting human arrogance.

    For a sympathetic critique of Bookchin's way of identifying variety and subjectivity as natural teloi, see Thomas W. Simon, "Varieties of Ecological Dialectics," *Environmental Ethics* 12 (1990), 211–31; esp. 223–5.
16. Since the text remains somewhat unclear on several of the following points, I have attempted to provide a coherent interpretation that elaborates certain underdeveloped ideas.
17. This high point is not an end point, however. Bookchin's notion of a telos (he prefers to use the term "directiveness" in his later works) is dynamic: Even the realization of a potential will still be accompanied by change. In *The Philosophy of Social Ecology* he writes, for example: "Dialectical naturalism thus does not terminate in a Hegelian absolute at the end of a cosmic developmental path, but rather advances the vision of an ever-increasing wholeness, fullness, and richness of differentiation and subjectivity" (30). It seems to me that the notion of a dynamic telos that is not a contradiction in terms needs to be further clarified.
18. In *The Modern Crisis* Bookchin writes: "Society . . . attains its 'truth,' its self-actualization, in the form of richly articulated, mutualistic networks of people based on community, roundedness of personality, diversity of stimuli and activities, an increasing wealth of experience, and a variety of tasks" ([Philadelphia, PA: New Society Publishers, 1986], 59).

In turn, freedom is also *made possible* by the existence of diversity (see Bookchin, *The Philosophy of Social Ecology*, 106, 129).

19. *The Ecology of Freedom* also seems to suggest that there is a functional relationship between the ends of consciousness and variety. The development of consciousness may, in addition to being an end in itself, have been a natural strategy to enhance diversity in the most effective way. Thus "what is authentically 'good' for us may very well not be a purely human desideratum but a natural one as well" (*EF* 342; see also 315). This speculation forms the basis of Bookchin's controversial suggestions that the natural world might *need* us and that we have a *responsibility* to enhance its variety. Cf: "We can try to reclaim our legitimacy as the fullness of mind in the natural world—as the rationality that *abets* natural diversity and integrates the workings of nature with an effectiveness, certainty, and directedness that is essentially incomplete in nonhuman nature" (*EF* 366).

For critiques of this view, see especially Robyn Eckersley, "Divining Evolution: The Ecological Ethics of Murray Bookchin," *Environmental Ethics* 11 (1989), 99–116; Michael E. Zimmerman, "Rethinking the Heidegger-Deep Ecology Relationship," *Environmental Ethics* 15 (1993), 195–224. For Bookchin's response to Eckersley, see Murray Bookchin, "Recovering Evolution: A Reply to Eckersley and Fox," *Environmental Ethics* 12 (1990), 253–74.

The essence of the critiques is twofold: (1) Critics fear that such a view of human responsibility will lead to the subjugation of nature, Bookchin's critique of domination notwithstanding; (2) Critics argue that "other species don't need any help in realizing their potential" (Zimmerman, "Rethinking the Heidegger-Deep Ecology Relationship," 223).

Personally, I am made uncomfortable by the presumptuous tone of Bookchin's statements on this point. Yet I believe the critiques are misdirected: (1) Fear of a "sliding scale" leading to domination may be realistic in general, but makes little sense in the context of Bookchin's ethic, which so explicitly and consistently pursues a middle road guided by ecologically informed reason; (2) Bookchin is not talking about helping specific species realize their individual potential, but rather about fostering an optimum *variety* of species, which would realize a larger evolutionary potential. Given the staggering reduction of species variety that is going on today, this does not strike me as an inappropriate form of "intervention" (see also Bookchin, "Recovering Evolution," 272–3). Nor does this imply that humans have a responsibility to touch *everything* (which would indeed be an "ecologically updated, anarchist version of the labor theory of value," as Michael Zimmerman suggests [223]). Bookchin's arguments are fully consistent with the idea of "letting things be" when they are in a position freely to realize their potential.

20. This definition is based on Aristotle, *Nicomachean Ethics*, trans. W.D. Ross (Oxford: Oxford University Press, 1925), 1106^{a}15–20.

21. I have used the following criteria to determine which terms occur in the role of virtues:

    1. traditional virtues that are cited with approval (e.g., hope);
    2. emotional and/or intellectual dispositions and related actions that are presented as contributing to human and/or natural teloi (e.g., attunement);
    3. emotional and/or intellectual dispositions and related actions that are presented as removing obstacles to human and/or natural teloi (e.g., rebelliousness);

4. emotional and/or intellectual dispositions and related actions constitutive of the human telos, the maintenance of which, under present social and psychological circumstances, requires a degree of effort (e.g., autonomy, creativity).

Synonyms are listed together. There is a certain degree of overlap between some other terms as certain virtues are constituents of others (e.g., caring and respect are constituents of love).

22. I have used the following criteria to determine which terms occur in the role of vices:

    1. traditional vices that are cited with disapproval (e.g., greed);
    2. emotional and/or intellectual dispositions and related actions that represent an avoidable absence of the conditions necessary for the achievement of human and/or natural teloi (e.g., repression);
    3. emotional and/or intellectual dispositions and related actions that are presented as undermining the conditions necessary for the achievement of human and/or natural teloi (e.g., domination);
    4. emotional and/or intellectual dispositions and related actions that are presented as distractions that keep us from seeking human and/or natural teloi (e.g., fatalism).

    Synonyms are listed together. There is a certain degree of overlap between some other terms as certain vices are constituents of others (e.g., authoritarianism is a constituent of mastery).
23. Among these terms are a good number of intellectual virtues and vices that, like the traditional virtue of prudence, are included for their contribution to the *moral* life. Intellectual virtues mentioned are ecological sensibility, holistic thinking (aspect of former), intellectual sophistication, organic knowledge, prudence, self-reflexivity, theoretical discrimination, utopian thinking, and wisdom. Intellectual vices mentioned are blindness, denial, ignorance, instrumental rationality, irrationality, lack of imagination, reductionism, self-deception, stupidity, and wishful thinking. (Again, there is a degree of overlap since some virtues and vices are constituents of others.)
24. There are many passages in *The Ecology of Freedom* that discuss how conceptions of virtue have changed over time. See, for example, 80, 100, 108, 144, 148, 153, 156, 298, 301.
25. For other explicit criticisms of Marx, see *EF* 65, 113.
26. E.g., "Knowledge . . . is the first step toward curing our social pathology of rule, just as self-knowledge in psychoanalytic practice is the first step toward curing a personal pathology of repression" (*EF* 349).
27. Much modern Christian social criticism has incorporated features of Marxist analysis, especially in the development of the notion of systemic evil. Still, the root of this evil is often traced back to individual human sinfulness.
28. Bookchin, *Remaking Society*, 24.
29. Bookchin considers Marx's theory only correct insofar as it describes capitalist society, not as a universal theory of historical development.
30. On moving "through Marx," see also Murray Bookchin, "Beyond Neo-Marxism," *Telos* 36 (Summer 1978), 5–28; and several essays in Murray Bookchin, *Toward an Ecological Society* (Montreal: Black Rose Books, 1980).
31. For a parallel critique, see Thomas W. Simon, "Varieties of Ecological Dialectics." Simon criticizes "the priority of ethics over politics" (214) in much recent

ecophilosophy and proposes a more balanced approach that incorporates many features of Bookchin's thought.

32. Note also: "The very terms 'base' and 'superstructure' are alien to the outlook that permeates this book. Reductionistic and simplistic, these terms tend to reflect naive views of a reality whose wealth of interactions defies overly schematic and mechanistic interpretations" (*EF* 351).
33. This is also true for Bookchin's other works. See especially *Toward an Ecological Society* and *The Rise of Urbanization and the Decline of Citizenship* (San Francisco: Sierra Club Books, 1987).
34. See also Bookchin, *The Rise of Urbanization and the Decline of Citizenship*, 228.
35. For a similar line of reasoning, see ibid., 287. For a sympathetic, though little developed, critique of Bookchin's theory of social change, see Timothy W. Luke, "Social Ecology as Critical Political Economy," *The Social Science Journal* 24 (1987); 303–15, esp. 314. Luke's proposal that social ecologists should pay more attention to "the role of the state" and "the means of mass mobilization" (314) in bringing about social change seems to stand in tension with Bookchin's views on the state and the ideal size of human communities. Since Bookchin insists that there be "no disjunction between ends and means" in direct action (346), Luke's proposal may not be practicable within the framework of Bookchin's social ecology.

Chapter 4 Four

THE VIRTUE ETHIC OF THOMAS BERRY

THOMAS BERRY, PASSIONIST PRIEST and historian of cultures, is widely known for his attempt to provide the modern world with a cosmology that is both consistent with modern science and powerful enough to guide human conduct in an ecological age. In this chapter I will highlight the virtue ethic that accompanies Thomas Berry's cosmology, an ethic both rich in its details and thought provoking in its overall structure.

Like Murray Bookchin, Thomas Berry develops his ideas against the backdrop of Aristotelian tradition, yet there are significant differences between the two authors. For one thing, Berry's immediate discussion partner is Thomas Aquinas, not Aristotle. The resulting virtue ethic bears the marks of Christian (more specifically, Roman Catholic) sensibilities, even if it cannot be described as Christian in any orthodox sense of the word. Moreover, Berry's ethic is of the "narrative" type, rooted and consciously presented in relation to an explicit story, whereas the ethic of social ecology draws on a more plainly conceptual scheme. One of the aims of this chapter is to illustrate these and other differences and thereby to underscore the diversity that marks ecological virtue discourse. Above all I will seek to display the richness of Berry's ethic, *next to* Bookchin's rich ethic, as something to celebrate and as an impetus for further reflection.

The focus of this chapter will be Thomas Berry's book *The Dream of the Earth*, a collection of essays on topics ranging from technology and economics to Native American spirituality and cosmology.[1] A single vision connects the essays, the vision of an ecological age as the next step in the unfolding story of the universe. This vision also underlies *The Universe Story*, a later work coauthored by Thomas Berry and mathematical cosmologist Brian Swimme, in which the authors portray the birth of the universe and the development of galaxies, the solar system, the planet Earth, and the history of its life, all gathered in a single narrative.[2] At certain points in the chapter I will draw on *The Universe Story* as well.[3]

Like Murray Bookchin, Thomas Berry does not present a distinct or consciously developed virtue ethic. In fact, the terms "virtue" and "vice" do not occur in *The Dream of the Earth*. Yet the text does contain a wide range of concrete virtues and vices, both traditional and innovative. Various vices are described by means of the term "attitude" (e.g., our "aggressive attitude towards the natural world" [*DE* 5]), and in general Berry sounds the warning that "if we do not change our attitude and activities, our children and grandchildren will live not only amid the ruins of the industrial world, but also amid the ruins of the natural world itself" (*DE* 177). This call to change "our attitude and activities" alerts us to the fact that Berry's ethic can be characterized as a virtue ethic. I will explore the distinct features and contributions of this ethic by focusing on its narrative context, its unique catalogue of virtues and vices, and Berry's views on the relationship between the life of virtue, our current ecological predicament, and the prospects for social change.

Virtue and the Story of the Universe

According to Thomas Berry, human beings have typically understood their own existence within a larger cosmological context. Exemplified particularly by the creation myths found in most religions, cosmologies have long provided humans with a sense of rootedness and direction. The "story of the universe," told by Thomas Berry and Brian Swimme, functions similarly as a creation myth for a modern age. This contemporary cosmological narrative also sets the scene for a distinctive virtue ethic.

In its basic form the narrative is fairly simple, told as a grand drama in which the universe and its parts are the actors:

> The story of the universe is the story of the emergence of a galactic system in which each new level of expression emerges through the urgency of self-transcendence. Hydrogen in the presence of some millions of degrees of heat emerges into helium. After the stars take shape as oceans of fire in the heavens, they go through a sequence of transformations. Some eventually explode into the stardust out of which the solar system and the earth take shape. Earth gives unique expression of itself in its rock and crystalline structures and in the variety and splendor of living forms, until humans appear as the moment in which the unfolding universe becomes conscious of itself. The human emerges not only as an earthling, but also as a worldling. We bear the universe in our beings as the universe bears us in its being. The two have a total presence to each other and to that deeper mystery out of which both the universe and ourselves have emerged. (*DE* 132)

The same drama is related in much greater detail in *The Universe Story*. In this section I will highlight those aspects of the universe story that are particularly relevant for virtue ethics, namely the numinous quality of the universe, its dynamism, the violence inherent in its development, its orderly structure and the concomitant functionalism. Several of these aspects overlap with Thomas Berry's own attempt to provide a concise overview of the main points of the story in his "Twelve Principles for Understanding the Universe and the Role of the Human in the Universe Process."[4]

Like Murray Bookchin and most other ecowriters, Thomas Berry presents us with a holistic worldview. For Berry the universe is a diversified whole. "We live in a *universe*," he writes, "*a single, if multiform, energy event*" (*DE* 45–6). Human beings are integral parts of this whole, related to everything else through a common origin (*DE* 21; *US* 5). Yet unlike Bookchin, Berry interprets the whole in religious terms. The single energy event, in all its multiform expressions of creativity, has numinous qualities. It inspires awe and, like a personal deity, is referred to as a *subject*. According to Berry the universe "reveals" itself and it has a "primordial intention," namely, "to produce variety in all things" (*DE* 134). Similarly, on a smaller scale, the planet Earth is presented as a "living organism" with "spirit and person qualities as well as . . . physical aspects" (*DE* 18), which "gives unique expression of itself" (*DE* 132) and "will not long endure being despised or ignored" (*DE* 119). Such personification is not just metaphorical language for Berry, but expresses his belief that subjectivity pervades all of reality as the inner dimension of things. Thus, unlike Murray Bookchin, Thomas Berry subscribes to a strong type of holism.

The strong holism of the universe story has significant implications for ethics. Berry treats the numinous energy event that is the unfolding universe as the most basic good. It provides the foundation for all value and moral direction, giving a teleological shape to Berry's ethic.[5] Moreover, the numinous energy event directly sprouts the concrete virtue of awe or reverence in humans keyed to cosmological development through their genetic and cultural heritages. "Even before we give expression to any intellectual statement about the natural world," writes Berry, "we stand in awe at the stars splashed in such prodigal display across the heavens, at the earth in its shaping of the seas" (*DE* 198). This reverential attitude can be considered the most cardinal of virtues in Berry's ethic.

As an energy event the universe is not only numinous, but also dynamic. Divine energy birthed the universe and remains a creative force through which the universe unfolds in a unique and irreversible process. Berry frequently uses the term "spontaneity" to refer to this dynamic impulse, a term he links to Confucian thought as well as to the creative Word of the Gospel of John (*DE* 196–7; *US* 189).[6] Within individual parts of the universe, whether atoms or humans, this creative spontaneity manifests itself in the form of urges to behave or change in a particular way. Overall, the concept of spontaneity provides an important hermeneutical key for understanding Thomas Berry's cosmology. It guides his interpretation of ancient creation stories as well as his reading of modern scientific accounts of the origin of the universe, thus providing a link between tradition and science. Moreover, for Berry the concept of spontaneity also connects creativity at a macrolevel to creativity at a microlevel, thus providing a unitary theory of development.

Thomas Berry's view of the universe as a spontaneous, dynamic event has ramifications for virtue ethics as well. It implies that the basic good that grounds ethics is itself in flux. The basic good is a matrix that constantly pushes toward a greater actualization of possibilities. It functions as a final cause, a moral telos in the Aristotelian-Thomistic sense, except that there is no fixed "final" point. (Here Berry's teleology resembles Murray Bookchin's.) The moral life can ultimately only take its bearings from the *consistency of direction* in which the actualization process unfolds. This direction can be followed through the channels of spontaneity. Berry argues that humans ought to listen to the spontaneities they experience in themselves, if only because they would be foolish to think that they could outsmart the wisdom of a creative process that has come into existence through billions of experiments (*DE* 177).

Such listening to spontaneities must be seen as a creative undertaking, since humans are cultural, creative beings. Their listening is not a matter of blind following, but would be more adequately described by the image of "chiming in." Ethics, then, is a matter of creative cooperation. As a consequence its focus is, in Berry's own words, "biocentric" rather than anthropocentric (*DE* 21, 161; *US* 257–8).

This ethic of creative cooperation, mediated through the experience of spontaneities, also involves critical awareness. There is much that can interfere with human sensibilities, especially when cultural coding opposes genetic coding (*DE* 202). Hence "we must respond critically with these spontaneities to assure their authentic expression" (*DE* 196). In any event, cultural interference is no excuse to disregard cosmically rooted spontaneities:

> We have ultimately no other source of guidance that possesses such inherent authenticity or which can function so effectively as a norm of reference in our actions. In earlier times these spontaneities were considered as revealing the natural law, the ultimate inner norm of guidance for human conduct, since they are the human phase of those instincts that enable a bird to build its nest, find its food, and discover its migratory route. Ultimately these instincts come from that mysterious source from where the universe itself came into being as articulated entities acting together in some ordered context. (*DE* 196)

On this model, virtue ethics becomes synonymous with the authentic expression of spontaneity, and the effort implied in the cultivation of virtue becomes synonymous with the critical task of cutting through obstacles to spontaneous expression, as well as with the creative task of responding in culturally and individually specific ways.

If, then, the numinous character of the energy event provides the foundation of value, its dynamism provides the channel through which humans can have access to this foundation. One could say that it gives formal shape to the moral life, since humans must always be open to this channel, no matter what content is being communicated. Such openness requires attentiveness, critical reflection, and creativity. In Thomas Berry's ethic, these basic attitudes function as cardinal virtues in addition to the reverence that is directly inspired by the numinous nature of the energy event. Each is a necessary condition for the cultivation of all other virtues.

Another important characteristic of the universe as it develops is its violence (*DE* 216). Supernovas explode, particles and bodies collide and destroy each other, and, on Earth, millions of species become extinct. According to Thomas Berry, such incidents express the tensions manifest in the curvature of space.[7] Yet ultimately violence does not have the last word. Nature may be "both benign and terrible," but it is "consistently creative in the larger patterns of her actions" (*DE* 67; *US* 264). In other words, the overall direction of development is toward creation and not destruction. This is the "good news" inherent in Berry's story.

The good news is not unqualified: Violence may not have the last word, but it is an *indispensable* condition for creative advancement. As most five-year-olds can tell you, humans would not have existed if the dinosaurs had not become extinct. Perfect peacefulness can never be. Even so, one can distinguish between the necessary violence of natural processes and the human-induced violence that serves no constructive purpose save to satisfy short-term exclusive interests (*DE* 54, 67). The creative energy event is violent, but this does not mean that all violence is meant to be.

These ideas also have significant implications for virtue ethics. First of all, if violence is a necessary part of reality, the focus of human moral vision should not be peace but rather what Berry calls "a *creative resolution of our present antagonisms*" (*DE* 219). And if, as some virtue ethicists would assert, vision shapes virtue,[8] then presumably Berry's cosmology provides criteria for judging the genuineness and priority of virtues. For example, the romanticism of the person who seeks to undo all modern development in order to establish a pristine and fully harmonious back-to-nature society cannot be considered virtuous. And peaceableness cannot be considered a more important virtue than justice.

A second ethical implication of Berry's interpretation of violence is the crucial role of discernment and the concomitant choice of fitting virtues in violent situations. Humans need to develop the ability to tell the difference between inappropriate, human-induced violence and the unavoidable violence of natural processes (*DE* 217).[9] The former should be opposed, which requires the cultivation of resistance (a form of courage). The latter should, to a degree, be allowed to happen since it is a condition for creative renewal. Under such tragic circumstances it may be more appropriate to cultivate the virtue of resignation. Still, Berry is not a misanthropist. He is aware of the suffering that nature can inflict on sentient beings and proposes that humans may defend themselves against this violence as long as they do not destroy nature's potential for creative renewal (*DE* 67). Overall, Berry's vision of creative transformation calls for a virtue ethic in which, through discernment, opposing virtues of resistance and resignation are held in a tensive balance.

Finally, Berry's theory of violence also allows for a meaningful concept of vice without requiring the postulate of a reality that is destined to be peaceful. In Berry's work, vices appear as distinctively human qualities that are morally distinguishable from the evils inflicted by natural processes. They could be characterized as *sins against creativity*: deliberate attitudes and actions that disrupt creative processes (which may involve the creative development of individual beings) for short-term, purely selfish ends. Such sins against creativity also include the deliberate refusal to search for creative resolutions of existing antagonisms. Thus what marks vicious behavior is not simply its results in terms of violence, but its lack of fit within the larger positive development of the universe—and this lack of fit is primarily a matter of misdirected human will.

In addition to being numinous, dynamic, and violent, the energy event also unfolds in an orderly fashion. According to Thomas Berry, the creativity of the universe operates within the parameters set by the curvature of the universe. *The Universe Story* gives a detailed account of what this means. First of all, the curvature of the universe ensures that development issues in a network of relations rather than in chaos:

> The expansive original energy keeps the universe from collapsing, while the gravitational attraction holds the component parts together and enables the universe to blossom. Thus the curvature of the universe is sufficiently closed to maintain a coherence of its various components and sufficiently open to allow for a continued creativity. (*US* 260)

Second, this tempered development follows fixed patterns:

> The cosmogenetic principle states that the evolution of the universe will be characterized by *differentiation*, *autopoiesis*, and *communion* throughout time and space and at every level of reality. (*US* 71)

At the level of the development of life, these patterns manifest themselves as genetic mutation, niche creation, and natural selection (*US* 125). As a result, we can speak of an ordered universe, with the understanding that the order refers to a dynamic process rather than a fixed reality.

Crucial implications for virtue ethics follow once again. Most fundamentally, moral action is only realistically possible in an ordered context, which allows for a measure of predictability on which all responsibility is predicated. Not only is moral action made possible within the ordered context of Thomas Berry's cosmology, it also acquires concrete content. Whereas the numinous quality of the energy event provides the foundation of value, and while the authentic expression of its spontaneity provides the form of the life of virtue, the order found in the creative unfolding of the universe provides material principles that can be expressed in the cultivation of concrete virtues. Differentiation, autopoiesis, and communion must be followed, safeguarded, and fostered (*US* 251). This is what it *means* to listen to spontaneities and to show respect for the mystery of the universe. This is what all genuine virtues are about, that is, following the orderly matrix of cosmic development, just as all vices involve willful opposition to this matrix.

So far I have focused on the order of the universe at the macrolevel of cosmic dimensions and time. At this level we can only speak of an ordered *process*. Yet at a microlevel of planetary dimensions and human perception of time, it is also possible to speak of order within and among the *products* of this process. Even if the universe as a whole cannot be understood as a fixed order of being, order exists at the level of its parts when conditions of wholeness and integrity are present, that is, when one can speak of flourishing. Such micro-order is always temporary and space-specific, since the overall creative process continues to unfold everywhere in unique ways. Yet from our human perception of time and space it can—depending on which entities are observed—appear quite stable and comprehensive. As long as we do not blow this understanding of order out of proportion or see it as eternally fixed, it can be quite useful, even essential for a proper life-orientation.

The flourishing that characterizes such temporary and space-specific order can be perceived at various levels. Most significant on the scale of human perception are the planetary, communal, and individual levels. Thomas Berry points to a flourishing that fits each case.

Berry is particularly struck by the uniqueness of the planet Earth in the universe. The distinctive features of our planet—its immensely rich variety of nonliving and living elements—suggest a unique way of flourishing. The Earth finds its fulfillment in "celebrating the joy of existence" (*US* 3) through the sheer extravagance of creaturely forms. In other words, the diversity of our planet is its order. This order is a current reality (though we are on our way to destroying it), as manifested especially by the great variety of lifeforms that exist today. Still, Berry also reminds us that from a macroperspective this order is but a moment in an ongoing and open-ended (but not arbitrary) process in which each biological period achieves its own fulfillment, to be transcended again by the next biological period (*DE* 220).

The Earth is covered with communities, and each of these has its own conditions for integrity and wholeness. When these conditions are met (which may be possible in a variety of ways), one can speak of a flourishing community, or communal order. Berry especially highlights the flourishing of bioregions, life communities that may include humans: "An integral functional order exists within every regional life community. This order is not an extrinsic imposition, but an interior bonding of the community" (*DE* 167). The conditions of flourishing at this level correspond with the principles of ecology. Yet these principles also relativize the idea of communal flourishing: An ecosystem can have an optimal way of functioning, but it is inherently dynamic as well. What is optimal at one point in time may not continue to be optimal (though such changes usually occur slowly).

Individual members of these life communities also have their own ways of flourishing. Within the context of communal order, members are enabled "to achieve that fullness of life expression that is proper to each" (*DE* 167). What this entails generally depends on the specialization of each member. For example, a bird flourishes in its flying, a tree in its blossoming (*US* 157). Humans, being less specialized than plants or other animals, can flourish in a greater variety of ways (*US* 157–8). To a certain extent human flourishing is conditioned by genetic structures, yet cultural coding also plays an important role. The human is "only imperfectly human" at birth and needs a long acculturation period "for arriving at a truly human maturity, for the learning of language, for initiation into the rituals whereby humans coordinate their own activities within the cosmological order, for adjusting to appropriate roles within the social order, for acquiring artistic skills, for learning the stories, the poetry, the songs and music of the community" (*US* 158). Consequently, human flourishing can take as many forms as there are cultures. Still, genetic coding ensures that human flourishing always answers to a certain structure. For example, all human flourishing involves the use of language and reasoning, ritual, role-identification, and creative expression.[10] Thus, even in the case of the nonspecialized human species, one can speak of an ordered way of being.

The various levels of flourishing are intimately related and cannot be fully understood apart from one another. For example, Berry affirms the insight of popular environmentalism that the well-being of the natural world is a precondition for human well-being (*DE* xv). Berry also goes further, claiming that human fulfillment can only occur as *part of* a larger process:

> Our individual being apart from the wider community of being is emptiness. Our individual self finds its most complete realization within our family self, our community self, our species self, our earthly self, and eventually our universe self. (*US* 268)

In other words, the very notion of human flourishing assumes immersion in the flourishing of more comprehensive levels of existence. Here Berry's strong holism shines through again. At the same time, however, human fulfillment itself becomes an expression of the fulfillment of the life communities to which humans belong (*DE* 168), for one cannot point to the flourishing of the community without pointing to the well-being of its specific members.

All of this is eminently relevant for understanding the virtue ethic that is implicit in Thomas Berry's work. The concept of flourishing is central to many traditional and modern virtue theories. Typically, the claim is that the cultivation of virtue leads to human flourishing and that human flourishing is in turn (partially) expressed through the life of virtue. Thomas Berry's cosmology provides a distinct interpretation of human flourishing, which is both more flexible (and hence less elitist) than is typically the case in Western teleological virtue traditions and, at the same time, defined enough to avoid a total relativism. A matching virtue ethic will be open to cultural diversity, allowing for the unique articulation of virtues and vices in each social setting without attempting to reduce all of these to a common denominator. At the same time such a virtue ethic will be able to accommodate the thesis that there are likely to be classes of virtues that match the various basic structures by means of which human flourishing is reached (e.g., virtues that guide human reason, language, social life, and ritual). Finally, by putting human flourishing in a context of communal and planetary flourishing, all again conceived in the larger dynamic context of the creative development of the universe, Berry's cosmology challenges definitions of human virtue that only take the human dimension into account and invites reconsideration of the ancient ideas of natural and cosmic virtues.

The notion of order, understood as flourishing, brings in its wake the ethically laden notion of functionalism. Where there is order it becomes possible to identify the unique contributions of the elements that constitute this order. One can speak of roles and functions within a dynamic network of right relations. As we read in *The Universe Story*, "Each member of the Earth community has its own proper role within the entire sequence of transformations that have given shape and identity to everything that exists" (*US* 268). The "proper role" of human beings is of particular interest to Thomas Berry. Even though human flourishing can take on a variety of forms, when it comes to the place of people in the larger order of things it turns out that humans have very specific functions to fulfill. Human flourishing has to take place within the parameters set by these functions, which are, in turn, determined by the flourishing of the larger whole.

What are these human functions? First of all, within the overall context of cosmic development, humans are meant to be the brains of the universe, the place where the universe becomes conscious of itself and reflects on itself. This is the culmination of the development toward increased subjectivity that has marked the universe from the very beginning.[11] Second, humans are meant to be the ultimate celebrators of the universe. Through ritual, dance, music, and all other art forms, they can and should consciously rejoice in the mystery that has unfolded in such rich and varied ways. Third, within concrete life communities, humans must also discover their proper role, yet this role differs in each case depending on how each community has evolved. We can think, for example, of the way in which certain ecosystems in the northeastern United States became dependent on periodic brush burning introduced by Native Americans. Thus humans have various "proper roles" to play, even if within and in addition to the execution of these functions there may be room for variation.

In terms of ethics, the functionalism inherent in Thomas Berry's cosmology implies a degree of determinism as well as specific moral obligations. Berry believes that humans are genetically predisposed to execute their proper functions. In *The Universe Story* we read: "While the human situation is definitively changed from this earlier period we remain genetically coded toward a mutually enhancing presence to the life of community that surrounds us" (*US* 264). Moreover, insofar as the larger flourishing context is the most basic good (which Berry believes), the fulfillment of functions that vitally contribute to this good becomes a moral imperative. Indeed, Berry uses terms like "obligation" and "responsibility" to articulate this moral imperative. He even speaks of an "expectation," as though the universe itself (personified) expects humans to play a certain role:

> As with all other earthly beings, we are expected to enter into this process [of the story of the universe] within those distinctive capacities for human understanding and appreciation that provide our human identity. We are expected to enter into the process, to honor the process, to accept the process as a sacred context for existence and meaning, not to violently seize upon the process or attempt to control it to the detriment of the process itself in its major modes of expression. (*US* 251–2)

This passage also indicates how humans can wrongly interpret their roles, or not care about them at all. Clearly, genes do not fully predestine humans to fulfill their roles, which is why a supplementary moral imperative makes sense in the first place. Here education also becomes important.[12]

Thomas Berry's functionalism, which combines genetic coding with moral imperative, raises interesting questions about the scope of freedom in the moral life. Genetic coding limits freedom, but does not preclude it altogether. For example, Berry assumes that we are genetically coded to celebrate the universe through ritual (*DE* 200), but the concrete content of our rituals is freely chosen. Moreover, by engaging in such rituals, for example, rituals that celebrate the cyclical processes of the natural world, we "establish our world of meaning, our sense of reality and value" (ibid.). Consequently, we would not even recognize freedom as a value if it were not for our participation in this genetically determined activity. And because we experience this participation as a positive good, we also *feel* free, in the sense that we do what we enjoy doing.

Nevertheless, the moral imperative implicit in Thomas Berry's functionalism does limit freedom. To be precise, as humans we are not free (i.e., allowed) to choose not to use our minds and creative capacities (nor could we really), nor are we free (i.e., allowed) to use them against our life communities (even though we can, as we have proven). But we can, within these parameters, still make many choices. Exactly because humans are meant to be the brains and celebrators of the universe, there is room for freedom and creativity, since rationality and creativity are inherently open-ended. And again, we will likely experience our use of these faculties not as an obligation, but as a joy.[13]

These considerations bring into view the full scope of Berry's virtue ethic. Traditionally, Aristotelian-Thomistic virtue ethics has been closely tied to functionalism. Virtues have been keyed to functions, the fulfillment of which should lead to happiness.[14] Functions, in turn, have been keyed to nature. Thomas Berry's cosmology invites a similar pattern of conceptual relations. Yet it is significant that "genetic coding" takes the place of "nature." Since genetic coding is a more precisely testable concept, this ensures less arbitrariness in the articulation of virtues than we find in much of our past, when nature often served as an ideological concept with very little basis in testable experience. The substitution of genetic coding for nature also ensures a greater degree of variability in the articulation of virtues, since genetic coding still leaves much undetermined. Thus the functionalism tied to Thomas Berry's cosmology implies a less restrictive virtue ethic than we have typically found in Western teleological traditions.

In the sweeping drama of the universe story, Thomas Berry presents us with a universe that is numinous, dynamic, violent, and ordered in such a way that at a macrolevel we see laws of development, and at a microlevel a variety of flourishing and functions. The implications of this narrative for virtue ethics are many. The universe story grounds the moral life and gives it form and direction. As an orderly context in which moral responsibility becomes meaningful, it clarifies the freedom as well as limits of human moral action. It provides a general vision for the life of virtue: the enhancement of differentiation, autopoiesis, and communion, as well as the creative resolution of antagonisms. And the universe story also suggests universal classes of human virtues (guiding human language, reason, ritual, social roles, and creativity), as well as a list of cardinal virtues (reverence, attentiveness, critical reflection, and creativity), while leaving room for cultural diversity in the cultivation of these virtues. Thus the story of the universe, as a narrative conceptual context, provides the interpretive key to the rich collection of specific virtues and vices that can be found in Thomas Berry's work and to which I will now turn.

Another Distinctive Catalogue

Thomas Berry refers to the universe story as a "functional cosmology," a narrative that effectively gives meaning and direction to human life. I have already highlighted some of the general implications of this story for virtue ethics, and now I will cull a catalogue of virtues and vices from the story, which I offer as a concrete opportunity to compare the moral sensibilities of Bookchin and Berry. Yet in the midst of all this analytical work we must keep in mind one thing: The virtue ethic of Thomas Berry is a narrative ethic. The universe story does not just "add" dramatic presentation to an ethic that could be fully understood as a set of principles with matching virtues and vices. Rather, principles, virtues, and vices draw their lifeblood from the story in which they are embedded.[15] Thus we would do well to keep the universe story in mind throughout this section also.

Just like Murray Bookchin, Thomas Berry does not himself offer a comprehensive catalogue of virtues and vices that require our attention in an ecological age. The following concepts (or derivative terms, depending on the grammatical context) occur in the role of virtues in *The Dream of the Earth*[16] (concepts most frequently mentioned and/or particularly stressed by the author are printed in italic type; terms used as synonyms by the author are listed together):

acceptance of limitations
anger/resentment (also as vice)
awe/reverence
benevolence
biocentrism
care
carefulness
commitment
compassion
concern
confidence (trust)
cooperation
courtesy
creativity
critical reflection
daring (courage)
discipline
disengagement
(from commercial life)

empathy
graciousness
gratitude
healing
holistic thinking
honor
hope
humility
interdependence
joyfulness
loyalty
mindfulness
modesty
naiveté (postcritical)
nurturing
participation/intimacy/communion
realism
responsibility
sacrifice
sensitivity/listening/attentiveness/evocative attitude
sharing
spontaneity
submission
wisdom

The following concepts occur in the role of vices in *The Dream of the Earth*:

aggressiveness
anthropocentrism
arrogance/pride/self-inflation
blasphemy
brashness
consumerism
contempt
control/domination/conquering attitude
cruelty
dependence (on modern business corporations)
distrust
exploitation
foolishness
fundamentalism/absolutist thinking
instrumentalism
irresponsibility
lack of discipline
manipulation/violence/forcing
mechanistic thinking
naiveté/oblivion/limited thinking/simplicity (i.e., simple-mindedness)
neglect
objectification
otherworldliness/aversion/distancing
resentment/rage
romanticism/idealism/utopianism
scientific attitudes
suppression
trivialization (of inner meaning)

This catalogue shows a significant preponderance of virtues (42) over vices (28), matching the generally positive ambience of Thomas Berry's work. Again we find traditional terms (e.g., reverence, care; arrogance, foolishness) next to terms that express a more modern ethic (e.g., biocentrism, spontaneity; anthropocentrism, manipulation). And although some terms belong to common ecological parlance (e.g., humility, aggressiveness), others reflect more distinctive features of Berry's thought (e.g., intimacy; trivialization). In the remainder of this section I will highlight how two features affect the way in which Thomas Berry understands many of these virtues and vices, namely: (1) the central role of religious sensibility in Berry's work; and (2) the ambivalent role of Thomistic themes.

We have already seen how in *The Dream of the Earth* the universe is portrayed as an immensely powerful, evolving event, with interiority as well as intentionality. Berry expects (i.e., predicts as well as wants) the human response to this universe to be religious in nature. Much of the virtue discourse of *The Dream of the Earth* thus takes on a religious tone.

Significantly, Berry describes the transformation from vice to virtue in language reminiscent of conversion experiences as well as mystical encounters with the divine:

> After the distancing a new intimacy; after the mechanistic a greater biological sensitivity; after damaging the earth a healing. We need only look at the surrounding universe in its more opaque material aspects—look at it, listen to it, feel and experience the full depths of its being. Suddenly its opaque quality, its resistance, falls away, and we enter into a world of mystery. What seemed so opaque and impenetrable suddenly becomes radiant with intelligibility, powerful beyond imagination. *(DE* 69)

Here we hear echoes of Augustine's conversion,[17] of Luther's "tower experience,"[18] of John of the Cross's passage from solitary darkness into the abandonment of love.[19] What this passage tells us is that virtues such as intimacy, sensitivity, and healing come in the wake of religious insight; they are, in a sense, like Paul's "fruit of the Spirit" (Gal. 5:22–3).

Once we "are drawn" toward reverence for the mystery of the earth *(DE* 17), we learn to see the events of the world in a new light. The entire book *The Dream of the Earth* is a chronicle of such renewed vision. Exploitation of the earth is actually blasphemy against a sacred reality (*DE* 105). Despite environmental destruction, we can place our hope in the earth, which "will guide us through the peril of the present" (*DE* 222). We must follow "the earth's demands" that we participate "in the grand liturgy of the universe" (*DE* 215). Our moral responsibility is "commensurate with the greater knowledge communicated to us [by the earth]" (*DE* 48) and requires that we submit to and humbly follow "the guidance of the larger community on which all life depends" (*DE* xiv, 168). Ultimately, the avenue of religious vision leads to self-sacrifice: Realizing that major life systems of the planet are closing down, we must be "willing to see the human diminish so that other lifeforms might flourish" (*DE* 212).[20]

Because his moral discourse is quintessentially religious, Thomas Berry interprets key virtues in quite a different manner than Murray Bookchin interprets these same virtues. For example, both authors stress the importance of sensitivity and spontaneity, and both see these attitudes modeled in tribal cultures.[21] Yet for Berry, sensitivity connotes submission to the larger community surrounding us, and spontaneity comes down to obediently following the directives of the universe as mediated through our genetic endowment. For Bookchin, by contrast, sensitivity and spontaneity belong to the sphere of human *freedom* (a word with a remarkably low profile in Berry's discourse).

As for the notions of submission and obedience, Bookchin writes unequivocally, "I hardly mean that we must surrender ourselves to a mythical 'Nature' . . . , a Nature that demands human awe and subservience" (*EF* 25). On various occasions he treats obedience as a counterfeit virtue, a dangerous euphemism designed to support hierarchical structures (*EF* 55, 105, 122, 254). This example serves as a reminder that we can only adequately understand a certain author's ecological virtue discourse when we take into account this author's more general viewpoints.[22] A catalogue alone never gives a full picture, and we would do well to use catalogues with appropriate reservation.[23]

Berry's virtue discourse does not boil down to *blind* religious obedience, however. His emphasis on human creativity illustrates how he sees the balance between universal direction and human initiative:

> Our own actions can be truly creative only when they are guided by these deeper spontaneities. Human administration of the universe in any comprehensive manner is far too great a task for any controlled process on the part of humans, just as the movement of the arm to pick up and drink a cup of tea would hardly be possible if we were required to consciously manipulate each movement of the eye, the arm, the nervous system, the oxygen, and the blood flow. Yet we do the act spontaneously, with extensive awareness and control. There is deliberation, but also spontaneity. (*DE* 48)

Creativity is thus a matter of cooperation, not of slavish following (in fact, the latter would violate what we sense the term creativity is all about). Concretely, this means that the place for human creativity is especially in the application and "inculturation" of universal directives (*DE* 65–8).[24]

This brings me to my second point regarding the influence of Thomistic tradition on the virtue discourse of *The Dream of the Earth*. Berry's ethic is not just generically "religious" in character, but more specifically influenced by Roman Catholic, Thomistic theology. In the previous section I have already touched on some general signs of this influence, especially with regard to the teleological and functionalist character of Berry's ethic. Now my aim is to show the role of Thomistic themes in the interpretation of specific vices and virtues, namely, anthropocentrism, instrumentalism, sensitivity, and participation.

Berry's attitude toward the writings of Thomas Aquinas is ambivalent. He cites "the divine ideas of Thomas Aquinas" as an example of human visionary moments (*DE* 197–8), yet he also sees the shortcomings of Aquinas's vision for our modem age. As a result, Berry can use the authority of Aquinas to support his criticism of anthropocentrism and instrumentalism, yet his understanding of sensitivity and participation would most likely cause the angelic doctor to turn over in his grave.

Just in case anyone thinks that Christian tradition supports the comfort and convenience of an anthropocentric viewpoint, Berry shatters the illusion with a direct quote from the *Summa Theologiae*. One creature alone, says Aquinas, could not adequately represent God's goodness; hence the incredible diversity of creation.[25] Berry's conclusion: "From this we could argue that the community of all the components of the planet Earth is primary in the divine intention" (*DE* 79). God is a biocentrist, and hence our anthropocentric habits have no valid foundation.

Concretely this means that we cannot be purely instrumentalist in our relations with the nonhuman world. Again Berry calls upon Aquinas to support his position:

> Saint Thomas dedicated his efforts in great part to defending the reality and goodness and efficacy intrinsic to the natural world. The natural world is not simply object, not simply a usable thing, not an inert mode of being awaiting its destiny to be manipulated by the divine or exploited by the human. (*DE* 81)

Thus, much as Aquinas appealed to authority in presenting his own innovative virtue ethic, Berry appeals to Aquinas to undercut the power of those pernicious vices that have a reputation of being "traditionally accepted." Both catch their readers by surprise.

Yet Berry is selective in his use of Aquinas. He draws the line when it comes to the static cosmology that undergirds the work of the great medieval thinker:

> The difficulty with this cosmology is that it presents the world simply as an ordered complex of beings that are ontologically related as an image of the divine. It does not present the world as a continuing process of emergence in which there is an inner organic bond of descent of each reality from an earlier reality. (*DE* 128)

Though there may be universal truths, they are not fixed in an order. Hence what we need is "human sensitivity in responding to the creative urgencies of a developing world" (*DE* 133), not the search for unchanging natural patterns that Aquinas had praised.

Aquinas's cosmology is not only static, but it also draws a sharp distinction between the physical and the spiritual. Here too Berry parts ways with the medieval master, criticizing him for "little appreciating, apparently, that . . . immersion into matter might also be considered as immersion into the divine" (*DE* 151). Participation in the physical dimensions of the universe, communion with the visible, audible, touchable, tasteable, and smellable world is a true virtue for Berry, something that makes us genuinely happy and reunites us with our Source. Aquinas, who argued that true communion is a rational affair between rational beings only, would have been very surprised indeed.[26]

Overall, then, Thomas Berry's virtue discourse may be deeply religious and significantly indebted to the heritage of Thomas Aquinas, but in the end Berry turns Aquinas on his head exactly to be faithful to what he sees as genuine religion: a dynamic, embodied human response to the emerging universe.[27] Hence the catalogue of virtues and vices in *The Dream of the Earth* differs significantly from the *Summa*'s "special ethics"[28]—at least as much as it differs from Bookchin's nonreligious catalogue.

Religion, Virtue, and Social Change

Religion also plays a central role in Berry's understanding of how we can move from an ecological crisis to an ecological society. Unlike Murray Bookchin, Thomas Berry very much resonates with Lynn White's analysis of the problem as a religious one, and Berry's solution—though unmistakably "Berryesque"—reveals significant parallels with White's solution.[29] For both authors the key lies in the interplay between religion and the cultivation of vices and virtues.

In *The Dream of the Earth*, social analysis takes the form of an investigation into religious pathology. For Berry, neither the vices of certain individuals or groups, nor economic systems are to blame for our modern predicament, but rather a tragically misconceived myth, a "vision of industrial progress which *more than any other single cause* has brought about the disintegration that is taking place throughout the entire planet" (*DE* xii, emphasis mine; see also 17 and *US* 218).[30] In *The Universe Story* it is named the myth of Wonderland. Because this myth was created in modern history, the roots of the environmental crisis are essentially modern. Berry recognizes that Neolithic and classical civilizations already brought about environmental destruction, but "in the larger perspective the damage was sustainable" (*DE* 7). He does not believe that these civilizations were already so fundamentally flawed that they contained the seeds for the modern destruction of the environment. Only with modern American spiritual traditions, rooted in a mixture of Christian millennianism and scientific beliefs, do we see the potential for radical destructiveness arise (*DE* 113–5). Thus Berry's social analysis, which—pace Lynn White—leads to a spiritual cause in modern history, differs substantially from Bookchin's analysis, which traces the roots of the crisis far back into history to a combination of human vices and harmful social institutions.

But how, we may ask, can a myth have become so damaging? Does this assumption not conceal the question of responsibility by elevating an ideology to the level of a historical agent? Must we not look *behind* the myth to certain individuals or groups who skillfully and manipulatively *use* it to further their own interests at the cost of the environment and other human beings? Here Berry surprises us. No vicious persons or groups are behind the myth of Wonderland, but rather the deeply human experience of revelation, the experience through which we generally connect with the holy powers of the universe:

> To bring about the closing down of the life systems of the planet on such an order of magnitude is obviously not something that originated yesterday or something that arises out of some trivial miscalculation, academic error, or ideology such as the Enlightenment, or even out of the industrial age itself. These are all symptoms and consequences of a vast turn in human consciousness that originated deep in the origins of the human process itself. It must indeed have been associated with those revelatory experiences that we consider the most profound experiences ever to take place in human intelligence, experiences so profound that we consider them to come from some divine reality. Our deepest convictions arise in this contact of the human with some ultimate mystery whence the universe itself is derived. (*DE* 38–9)

What went wrong is that our natural predilection for mystical vision got sidetracked from genuine spiritual forces and hitched onto false substitutes. This happened when, under the influence of eighteenth-century science, we began to see the world as "objective" and manipulable, as a world we could fabricate (*DE* 39–41, 115).

The resulting myth of Wonderland is powerfully seductive, exactly because it occupies those channels from whence humans have always been empowered.[31] People come under its spell and end up in a "trance state" (*DE* 119, 126) in which their listening and thinking capacities are compromised:

> We have been away somewhere, entranced with our industrial world of wires and wheels, concrete and steel, and our unending highways, where we race back and forth in continual frenzy. (*DE* 1)

The best way to describe this state is in terms of pathology. Berry uses images such as autism (*DE* 16) and especially addiction (*DE* 148) to indicate both the character and the seriousness of the situation.

By introducing the concept of pathology in his social analysis, Berry steers away from the question of *moral* responsibility. Just as addiction to alcohol or drugs cannot be counted as vicious (though it used to be), so the entrancement created by the myth of Wonderland is not a vice.[32] People who are so entranced cannot help themselves. Nor can they do anything to get out of their predicament, for they typically deny the seriousness of their state (*US* 254). Only a crisis could break the spell:

> As with any addiction, it could only become worse. The pathology was too deep and too universal to eradicate at the time. It could only continue

> until the bombs were dropped on Hiroshima and Nagasaki and industrial nations began construction of nuclear warheads sufficient to extinguish all the greater lifeforms on the planet. (*DE* 148)

Thus neither the cause of the ecological crisis nor the first initiative toward its solution is a matter of moral responsibility. Vice and virtue did not play a role up to the point of crisis.

But they do play a role now that we are waking up to the shock of ecological disaster. Recognizing that our disease draws on sacred powers, our only hope is a cure that draws on these same powers, but in a wholesome rather than distorted way. This means that we must learn to regain access to our instinctive resources, we must learn to shut out the chatter of our corrupted brains: "What is needed is not transcendence but 'inscendence,' not the brain but the gene" (*DE* 208). Such a change of orientation requires practice and persistence, as our addictive habits are deeply engrained. Thus the cradle of social transformation turns out to be a personal virtue ethic, an ethic that prepares us for the influx of the sacred (see *DE* 211).

But a strange virtue ethic this is indeed. If in the past we sought to overcome social evils by letting the brain rule our instinct, now instinct is to redirect our brain! If in the past the transformative power of the divine was made manifest through the arduous work of the intellect (Thomas Aquinas's virtue theory, for example, is predicated on this natural law assumption), now our embodied selves are the source of revelation, and our effort consists exactly in not making too much of an effort.[33] Social change is no longer a matter of control but of letting go.

Then the reenchantment will happen. We will again see the Earth as a living reality (*DE* 21) and appreciate the sacredness in each part of the universe. The new story will come alive for us, and we will find ourselves energized, committed to make the requisite changes (*DE* 46). Thus, if a distortion of sacred powers lies at the root of the current crisis, a distortion ultimately beyond the realm of moral culpability, a new religious orientation will bring social and ecological healing (*DE* 87; *US* 250), an orientation made possible through active passivity, through the cultivation of spontaneity.

Once we partake of this vision and experience its energizing power, a whole new range of virtues becomes both necessary and possible. These are what I have earlier called the "fruit of the Spirit," and they not only lead to greater personal fulfillment, but also become the basis for social transformation. Thus the "letting go" issues into a new discipline, guided not by the rule of the mind but by the persuasive, resonating power of the new Story (*DE* 124).

Overall, Berry's vision of the dynamics of social change is captured by the basic principle, articulated in *The Universe Story*, that "we cannot change the outer world without also changing our inner world" (*US* 250). Again we hear the echo of Lynn White. The emphasis is on changing the inner world first; the rest will follow. Social transformation is mostly left to the dynamics of religious conversion and the cultivation of new virtues. Nor does Berry put much of his hope for social transformation on human reason: The emergence of an ecological society depends on the somewhat paradoxical task of cultivating the intuitive.[34] Ultimately the possibility of change, including social change, belongs to the mysteries of the universe.

CONCLUSION

The virtue ethic of *The Dream of the Earth* is in many ways worlds apart from the virtue ethic of *The Ecology of Freedom*. After taking a close-up look at both, we cannot but appreciate the diversity that marks ecological virtue discourse. Yet if we only emphasize differences, we overlook the real and potential symbiosis of these approaches. For the greater the variety of moral dialects, the greater the chances that a wide variety of people will find a language that appeals to them, a language that will affect their hearts and their hands and empower them to usher in, as Berry would put it, the Ecological Age.

NOTES

1. Thomas Berry, *The Dream of the Earth* (San Francisco: Sierra Club Books, 1988).
2. Brian Swimme and Thomas Berry, *The Universe Story: From the Primordial Flaring Forth to the Ecozoic Era* (San Francisco: HarperSanFrancisco, 1992).
3. Page references to *The Dream of the Earth* and *The Universe Story* will occur in parentheses in the text, preceded by *DE* or *US*, respectively.
4. Thomas Berry, "Twelve Principles for Understanding the Universe and the Role of the Human in the Universe Process," in *Thomas Berry and the New Cosmology*, eds. Anne Lonergan and Caroline Richards (Mystic, CT: 23rd Publications, 1987).
5. Insofar as this good is numinous and of cosmic dimensions, the teleology of Berry's ethic is more Thomistic than Aristotelian.
6. Berry studied Chinese as well as the classics of Chinese thought and taught at the Center for Asian Studies at St. John's University in New York from 1960–66.
7. More specifically, in *The Universe Story* violence is traced to the resistance of matter, the energy cost of creating order, and "the tendency of all things toward fulfillment of their inner nature" (52–3).
8. See especially Stanley Hauerwas, *Vision and Virtue* (Notre Dame, IN: Fides Press, 1974).
9. The distinction between inappropriate, human-induced violence and natural violence

parallels Thomas Aquinas's distinction between moral evil (*malum culpae*) and evil suffered (*malum poenae*) (*Summa Theologiae* I.48.5, trans. the Fathers of the English Dominican Province [New York: Benziger Brothers, 1948]).

10. Berry bases this claim on the findings of physical anthropology (see *US* 158).
11. There is a tension in Berry's work between the idea that the universe develops in an open-ended way toward ever greater subjectivity and the idea that we have, with the human, reached a culmination point. Passages that suggest the latter at times also imply that this culmination point was a cosmic aim from the very beginning, a telos in the traditional Thomistic sense.
12. See *US* 256: "The primary purpose of education should be to enable individual humans to fulfill their proper role in this larger pattern of meaning."
13. One is reminded of Augustine's and Luther's expositions on the freedom of the Christian life.
14. For an analysis and critique of Aristotelian functionalism, see Susan Moller Okin, *Women in Western Political Thought* (Princeton: Princeton University Press, 1979), chap. 4.
15. The priority of narrative over principles and virtues, a key assumption in narrative ethics, also fits Berry's own claim that humans are wired to derive meaning and direction from narratives. Having a functional cosmology is one of those universal structures through which human cultural coding expresses itself. Hence the pervasive occurrence of creation myths across times and cultures.

    Interestingly, while the introduction of a dynamic universe tests the limits of Aristotelian-style teleological ethics, Berry (who stands in this tradition) jumps the limit and lands on the new possibility opened up by the very introduction of dynamism: the possibility of narrative ethics. What is added is a specific way of deriving direction: not from a final cause, but from a meaningfully unfolding story. From the vision provided by this story, all else follows. Nonhuman nature in its macro-, meso-, and microdimensions can be interpreted and made morally operative, and human life can now be seen as more or less fitting within the trends of the developing story.
16. For the criteria used to determine which terms occur as virtues or vices, see chapter 3, notes 21 and 22.
17. "It was as though the light of confidence flooded into my heart and all the darkness of doubt was dispelled," writes Augustine in his *Confessions* (New York: Penguin Classics, 1961), 178.
18. "As I had hated the phrase 'the righteousness of God' before [thinking it referred to divine judgment], I now valued it with equal love," writes Luther, reflecting on his conversion experience (Gerhard Ebeling, *Luther: An Introduction to His Thought*, trans. R. A. Wilson [Philadelphia, PA: Fortress Press, 1970], 40).
19. There are differences too: Berry criticizes the otherworldliness that has marked the conversion experiences and mystical encounters of so many Christians, since it discourages concern for the natural world (*DE* 132).
20. Berry is careful in his use of the notion of sacrifice. In *The Universe Story* we read: "The word *sacrifice* may be too encrusted with a history of misuse to serve us in our struggle to understand what is required of us. But for our ancestors a sacrificial act was a way of making holy, especially when the act had a bitter dimension to it" (59), and also, "The pursuit of pain for its own sake is as pathological as a life dedicated to avoiding pain of any sort" (60).
21. For tribal models in *The Dream of the Earth*, see esp. 4, 14, 109, 190–1.

22. Insofar as theory is influenced by practice, this in turn may require that we learn about the author's social context.
23. Rather telling in this context is also how differently Berry and Bookchin value the role of the traditional shaman. Bookchin criticizes the shaman for embodying a hierarchical mind-set, whereas Berry sees the shaman as a prime model that can help us in the cultivation of ecological awareness. He writes: "The next stage of scientific development will require above all the insight of shamanic powers, for only with these powers can the story of the universe be told in the true depth of its meaning" (*US* 238).
24. I see a strong parallel here with how the Roman Catholic magisterium has recently delineated the role of moral theologians in the church.
25. According to the full text cited by Berry, "because the divine goodness 'could not be adequately represented by one creature alone, he produced many and diverse creatures, that what was wanting to one in the representation of the divine goodness might be supplied by another. For goodness, which in god is simple and uniform, in creatures is manifold and divided; and hence the whole universe together participates the divine goodness more perfectly, and represents it better than any single creature whatever'" (*DE* 79; Berry cites *Summa Theologiae* I.47.1).
26. See *Summa Theologiae* II–II.24.1 and II–II.25.3; Aquinas also argues that "the inclination of our appetite to visible goods" inhibits our love of God, who transcends the sensible world (II–II.24.2 ad 2).
27. Berry himself may not draw such a strong conclusion, since he believes that Aquinas has a "sense of the sacred character of the natural world as our primary revelation of the divine" (*DE* 81). Personally, I think that in this case Berry stretches Aquinas's thought further than his writings warrant.
28. I.e. *Summa Theologiae* II–II.
29. For a discussion of White's thesis in *The Dream of the Earth*, see 116.
30. In *The Universe Story* the crisis is also attributed to a failure to understand: "The failure of humans to understand this change of consciousness [from an abiding cosmos to an ever transforming cosmogenesis], and their corresponding failure to integrate human technologies with the technologies of the natural world, are perhaps the most profound cause for the disturbed condition of the planet in this late twentieth century" (223–4). This need not imply an inconsistency in Berry's thought: The failure to understand could be the result of the industrial myth's power.
31. Hence Berry can observe: "That consumption has something sacred about it is obvious from the central position it now occupies" (*DE* 115).
32. Berry does hint, however, that corporations have recently started a deliberate exploitation of our mythical inclinations through the "corporate culture" movement (*DE* 55–6).
33. Still, Berry's work remains true to another fundamental natural law premise, namely, that human behavior must fit within its natural context. Thus he writes, "Our own presence to the universe depends on our human identity with the entire cosmic process" (*DE* 17).
34. This paradox parallels the difficulties faced by medieval monks when they tried to cultivate the virtue of humility. The more they succeeded, the more they risked becoming proud of their achievement, thus losing the very humility they had attained. See J. Philip Wogaman, *Christian Ethics: A Historical Introduction*, Louisville, KY: Westminster/John Knox Press, 1993, 65.

Part 2 Two

ECOLOGICAL VIRTUE THEORY

Chapter 5 Five

FOUR ARISTOTELIAN CRITERIA FOR AN ECOLOGICAL VIRTUE ETHIC

THE FIRST PART OF this book has been mostly expository. So far, my aim has been to show that a rather rich and diverse virtue discourse is emerging from the growing body of ecoliterature. Ultimately, however, I am not interested in just describing a new phenomenon. Ecological virtue discourse is emerging because people are searching for more ecological ways of living. Practical goals undergird this language, and this means that exactly its richness and diversity raise some practical problems that cannot be ignored. In the thick maze of variety we may find our niche, but we can also get lost. Then we need to be able to get back to some basic criteria, some clear reminders of what the whole exploration is all about. In other words, a bit of theory can be very practical. In the remaining chapters I will construct elements of an ecological virtue theory—elements, not a complete theory, as practical questions tend to arise in a haphazard manner and it seems more worthwhile to stay close to the practical questions than to risk losing touch with reality for the sake of systematizing (much as I admit loving the latter).

I do not see the process of constructing ethical theory as a *creatio ex nihilo*, but rather as a creative correlation of existing moral resources. In the process of correlating, critical and new insights emerge. The following chapters bring together ideas gathered from ecoliterature, from Western moral traditions, from women's experiences, from psychological reflection, and from human contemplation of the nonhuman world. Sometimes these ideas clash, sometimes they complement each other, sometimes they provide mutual reinforcement. In the midst of it all, I will critically select and connect, weaving threads of theory that I hope will have practical value as we search to live out an ecological virtue ethic.

The first practical question is: Does the richness and variety of ecological virtue language mean that potentially "anything goes" when it comes to cultivating virtues, or are there limits? My own sense is that there are indeed some limits. In this chapter I will propose four: boundary conditions that can help us see the parameters of the life of virtue in an ecological age. I will show how such concrete criteria can be derived from contemporary ecological discourse, but I will also justify them in the context of an established philosophical tradition. By following this method I hope to demonstrate that we can honor the wisdom that inheres in lived ecological praxis and yet maintain a critical perspective on this praxis by means of the wisdom that inheres in tradition. Thus we can increase the chances that the criteria we end up using are both practicable and trustworthy.

As in chapter 3, I will use Murray Bookchin's *The Ecology of Freedom* as my sample of actual discourse.[1] This text is particularly appropriate because Bookchin exercises a critical selectivity that witnesses to the—albeit largely implicit—operation of a good number of criteria that guide his understanding of what makes for genuine virtues. For the purposes of this chapter I will limit myself to a group of four criteria that take the form of psychological boundary conditions of the life of virtue. I will show how such conditions can be systematically culled from the text by paying close attention to those attitudes the author explicitly marks as not genuinely virtuous.

As a sample of tradition, I have chosen the virtue theory of Aristotle. Thus I hope to honor the recent upsurge of interest in Aristotelian virtue ethics among philosophers. This chapter will be of particular interest to those who seek to adapt that ancient tradition to the new situations and needs of an ecological age. In addition, my choice of Aristotle was inspired by Bookchin's own respect for this Greek philosopher. Subjecting the criteria operative in Bookchin's work to the critical light of Aristotelian reasoning seemed like a fair undertaking. This chapter will also be of interest to those who seek to substantiate the insights of social ecology by means of systematic reflection informed by one of the traditions that feeds into social ecology.

That Murray Bookchin exercises a careful and reasoned selectivity becomes very evident when we pay attention to what he *does not* consider virtuous, despite its virtuous appearance. Here I do not have in mind those traditional terms that never receive mention in *The Ecology of Freedom*—after all, it could just be coincidence that traits such as compassion, honesty, loyalty, cowardice, gluttony, and lying are not discussed. Rather, I have in mind (1) those traditional terms that Bookchin explicitly rejects as being falsely considered to be representative of virtuous or vicious dispositions and actions. I will refer to these terms and their referents as *counterfeits*[2] (for obvious reasons they do not appear in the catalogue in chapter 3, but they do occur in the text). And (2) those traditional *interpretations* of virtues that Bookchin explicitly rejects as failing to describe the essence of those virtues; I will refer to these interpretations as *semblances*.[3]

Distinctions between counterfeits, semblances, and genuine virtues and vices can be very telling, for they express criteria that define a particular understanding of the life of virtue. Bookchin's penchant for making precise theoretical distinctions as well as his characteristically anarchist suspicion[4] (and perhaps also his strident style) have yielded an unusually large number of counterfeits and semblances in comparison to what one normally finds in ecoliterature.[5] From these we can glean criteria that are indicative of the life of virtue as it appears from the perspective of social ecology and that may also serve the development of virtue theory in general.

In the remainder of this section I will analyze Bookchin's discussion of the counterfeit vice of sensuousness, as well as passages that identify semblances of the virtues of love, restraint, honor (having a sense of personal moral probity), and personality (having a sense of self-cultivation). Together these discussions yield four general insights, all of which draw links between the capacity for moral discernment and (socially determined) psychological conditions. I will show that these insights can be corroborated in light of traditional Aristotelian virtue theory so that they, in effect, become four psychological boundary conditions that can function as negative criteria in the process of moral discernment, showing us which judgments to distrust.

THE REPRESSION TEST

In *The Ecology of Freedom*, sensuousness is treated as a counterfeit vice, that is, as a commonly misjudged disposition that should be liberated from its negative stigma. Sensuousness has long been suspect in Western tradition. It implies a dwelling on the impressions of the senses, especially the sense of touch, *for its own sake*. This was definitely unacceptable to Christian theologians who typically held that our enjoyment of the senses should always point to God. Even Aristotle singles out the sense of touch as particularly in need of watching, lest we turn into brutes or animals.[6] It is this long tradition that Murray Bookchin reacts to when he sneeringly speaks of "the 'vices' of sensuousness" (8). In Bookchin's analysis, sensuousness is perceived as vicious by a repressive psychic structure that results from the "discipline of rule" that marks hierarchical societies (8).[7] Bookchin himself sees sensuousness as a positive trait and praises the culture of the 1960s for finally acknowledging its value. He suggests that the cultivation of sensuousness may even have revolutionary potential, for an integrated ecological society can only be built by integrated, embodied persons. He does not fail to note that "what marked the great utopians was not their lack of realism but their sensuousness, their passion for the concrete" (325).

If there is a general, experientially based insight implicit in Bookchin's treatment of sensuousness, it would be that *the judgment of a repressed psyche is not a good standard for determining what is virtuous.*[8] This insight is analogous in *structure* (though not in content) to the traditional Aristotelian warning that reason under the influence of inappropriate levels of fear or desire is not a good standard for determining what is virtuous.[9] I will take this structural correspondence as a starting point to build an Aristotelian justification for the general insight that I have drawn from Bookchin's text.

Prudence, writes Aristotle, includes the ability to perceive what is good (in which it is distinguished from the simple calculation of the best means to achieve a particular end).[10] This ability can become distorted under the influence of desire or fear. For example, excessive sexual desire can lead a person to confuse experiential desiderata with unchanging goods, so that he or she becomes intemperate in giving in to desire (while believing all along that he or she is temperate!). Or excessive fear can lead a person to see his or her own survival in a battle as the ultimate good, forgetting the larger good of defending the community. Since prudence is corrupted by inappropriate levels of fear or desire, we could say that these levels function as negative boundary conditions for the life of virtue in Aristotle's ethic. In other words, whenever such levels are detected, caution rather than trust is warranted.[11] This is especially important since there are no external standards by which we can determine what is virtuous in a particular situation. Only prudence can make that judgment, and it is crucial that prudence itself not be corrupted.

Along similar lines, and remembering Murray Bookchin's suggestion that a repressive psyche wrongly classifies sensuousness as a vice, we can also argue that psychological repression compromises the exercise of prudence and hence represents a negative boundary condition of the life of virtue. After all, repressed persons *uncritically confuse* that which is repressed (of which they are unaware) with evil. Such persons are not free to consider the objects of their repression as potential moral goods, because the repression is the result of a preexisting, unbearable belief that something is shameful or bad. Yet because they are not aware of this lack of freedom and are made profoundly uncomfortable by reminders of the repressed object, they will still confidently, stridently, and even defensively pronounce as vicious or evil whatever they have repressed (likely claiming such ideologies as "the voice of conscience" or "the law of nature" in support). If prudence includes the ability to perceive what is good, we are well advised to mistrust the judgments of a repressed psyche, particularly in those situations when such a psyche declares things to be evil or vicious. I will call this negative boundary condition the *repression test.*

The Alienation Test

Just as Bookchin's treatment of sensuousness as a counterfeit vice suggests a psychological boundary condition for the life of virtue, his identification of semblances of the virtues of love, restraint, honor, and personality points toward psychological criteria that can serve to safeguard the exercise of prudence.

Whereas counterfeits are terms that should either not be used at all (Bookchin seems to have no use for reverence, obedience, charity, and simplicity[12]) or that should only be used if we are willing and able to reverse their value (e.g., from negative to positive in the case of sensuousness), semblances are mistaken interpretations of terms that have a legitimate use. The correction of these mistakes does not require a reversal of value. In written sources, the misinterpreted terms often occur in quotation marks, an author's way of alerting readers to the presence of a distortion. As in the case of counterfeits, the presence of semblances in a text presupposes that the author has a clear view about what distinguishes the "mistaken" from the "correct." Through close reading, we can tease out the operative criteria. In doing so, we both clarify the author's view of the life of virtue and collect resources for the creative task of adjusting existing virtue theory.

A second criterion, which I will call the *alienation test*, can be gleaned from Bookchin's discussion of a particular semblance of love. The virtue of love plays an important role in Bookchin's social ethic. Love, described as a natural as well as a cultural attribute, enables us to see the other as an active subject like ourselves, rather than as an alien and passive object. Thus love is the basis of a nonhierarchical, nonmanipulative sensibility. Yet Bookchin issues a warning against extending the virtue of love to nature, as is so frequently done in ecoliterature:

> Expressions like "love of nature" . . . are permeated by the problematical relationships of our own society and mentality. Preliterate humans did not have to "love" nature; they lived in a kinship relationship with it, a relationship more primary than our use of the term *love*. (49)

This passage suggests that our use of the term love in connection with nature falls short because it is more a sign of our alienation than of a genuine connectedness. One could say that it expresses the *wish* for a virtuous relationship rather than the cultivation of the virtue itself. And while the wish may be in the right direction to the extent that it expresses a desire for a more primary relationship, its manifestation in our lives should not be confused with evidence of virtuousness. For example, our frequent hiking trips, our ecstatic reports of natural beauty, and our declarations of love for the natural world may at root only express a desperate wish to be more connected to nature; the level of our enthusiasm may ironically be a measure of the degree of alienation and insecurity we feel, since it is through our enthusiasm that we try to deny the truth to ourselves and to others. Thus we continue to pursue what we think are virtuous characteristics, not recognizing our failure nor the misplaced nature of our efforts.

Bookchin's treatment of love of nature as a semblance of the virtue of love suggests a second experientially based general insight, namely, that *the judgment of an alienated consciousness is not a good standard for determining what is virtuous.* Again this insight can be justified in light of the traditional assumption that prudence is corruptible and needs safeguarding. The exercise of prudence not only involves the ability to perceive what is good, but also the ability to judge which means are appropriate to achieve a certain end in the context of a particular situation. An alienated consciousness is unable to select appropriate means *because it does not accurately perceive the situation.* Alienation by definition implies a lack of connection with one's situation, which leads to various mistakes of judgment, including moral mistakes. The alienated hiker who is declaring love for nature does not realize that the situation first calls for a solution to "the problematical relationships of our own society and mentality," which will create the possibility of a more primary relationship with nature in which appropriate virtues such as ecological sensibility and solidarity can be cultivated. At that point it would become clear that the alienated attempt to cultivate love in relation to nature was like trying one's wit at a funeral: One wants the scene to be happier, but this is not the way to go about it (and no matter how funny the jokes, they are not even genuinely witty in such a setting, just as the hiker's love would only be a semblance).[13] In general, then, we should mistrust moral judgments born from alienation, just as we should be weary of repression. Like the repression test, the alienation test functions as a boundary condition for the life of virtue.

The Guilt Test

The Ecology of Freedom also shows how semblances may be generated by an unresolved sense of guilt, suggesting a third psychological criterion which I will call the *guilt test.* The interpretation of the virtue of restraint championed by some environmentalists provides a case in point. Although Bookchin believes there is a role for restraint (or temperance) in the moral life, he makes it very clear that restraint in our relations to nonhuman nature does not mean a complete hands-off approach:

> Today, when the hunter-gatherer's mere parasitism of the environment has emerged as a virtue in juxtaposition to contemporary man's insane exploitation, we tend to fetishize restraint to the point of passivity and

> nondoing. Yet the matricentric horticulturist managed to touch the earth and change it, but with a grace, delicacy, and feeling that may be regarded as evolution's own harvest. (61)

The virtue of restraint, then, is a matter of delicate interaction and as such represents the "golden mean" between passivity and exploitation. Yet a guilty modern conscience, eager to expiate the exploitative mistakes of the past, confuses the virtue of restraint with the related vice of deficiency, namely, passivity. Thus arises a semblance which, by reason of its extremism, is bound to be harmful.

Embedded in Bookchin's comments on restraint is the general insight that *the judgment of a guilty conscience is not a good standard for determining what is virtuous.* Once again the traditional understanding of prudence and its corruptibility substantiates this insight. A guilty conscience interferes with the operation of prudence *because it destroys the sense of the mean.* This is true irrespective of whether the feeling of guilt is justified or not. A person who acknowledges a moral mistake, who feels bad about it, and who wants to make amends, is likely to overcompensate and swing from one extreme to the other without realizing that they have missed the desirable mean. For this reason certain moral judgments of a guilty conscience are not to be trusted. The guilt test thus functions as a third psychological boundary condition for the life of virtue.

THE FETISHISM TEST

Finally, *The Ecology of Freedom* provides two examples that indicate how our capacity for moral judgment can also be compromised by the dynamics of fetishism. These examples suggest a fourth psychological criterion that I will call the *fetishism test.* According to Bookchin, traditional ideals and virtues sometimes get swept up by the fixations of a certain culture and acquire an entirely new meaning, even as the old terms continue to be used. If we are to follow Bookchin's Marxist-inspired analysis, this is what has happened to those virtues that could be called honor[14] (having a sense of moral probity) and personality (having a sense of self-cultivation). These virtues have become victims of the commodity fetishization that has accompanied the emergence of the capitalist market system:

> The restraints that Greek social theorists like Aristotle tried to place on the market . . . were completely removed, and objects or use-values began to infiltrate the lofty human goals that society had elaborated from the days of their conception in the *polis*. The ideals of the past, in effect, had become so thoroughly bewitched by things that they were soon to become things rather than ideals. Honor, today, is more important as a credit rating

> than a sense of moral probity; personality is the sum of one's possessions and holdings rather than a sense of self-awareness and self-cultivation. (71)

In each case, a fetish has become the measure of virtue, giving rise to rather shallow and subtly pernicious semblances of honor and personality.

This passage suggests the following general insight: *The judgment of a consciousness governed by fetishism is not a good standard for determining what is virtuous.* Again the problem is the corruption of prudence. Where our reasoning capacity would normally judge both the meaning and the achievement of virtue in light of a telos (and the passage implies that "a sense of moral probity" and "a sense of self-awareness and self-cultivation" would pass this test), its judgment is deflected by the influence of a fetish that usurps the place of the telos. Because this usurpation occurs without any deliberation about moral ends, we cease to operate as moral agents and become slaves to the fetish; hence the importance of the fetishism test as a fourth psychological boundary condition for the life of virtue.

Conclusion

The counterfeit vice of sensuousness and semblances of the virtues of love, restraint, honor, and personality very concretely illustrate Murray Bookchin's vision of the life of virtue. Just how concrete this ethic is becomes clear when we consider how easy it is to disagree with Bookchin's specific judgments, such as his insistence that we are mistaken to try and love nature or take a hands-off approach in dealing with it. Yet irrespective of whether we agree or disagree with the concrete embodiment of social ecology's virtue ethic, the arguments on which Murray Bookchin bases his distinctions between specific counterfeits, semblances, and genuine vices and virtues suggest general insights that can be independently corroborated. I have shown how, through close interpretation, one can glean these general insights from the text and subsequently test them in light of traditional Aristotelian virtue theory. As a result of this process I have proposed four criteria that take the form of psychological boundary conditions for the life of virtue:[15]

1. The repression test: The judgment of a repressed psyche is not a good standard for determining what is virtuous;
2. The alienation test: the judgment of an alienated consciousness is not a good standard for determining what is virtuous;
3. The guilt test: the judgment of a guilty conscience is not a good standard for determining what is virtuous; and
4. The fetishism test: the judgment of a consciousness governed by fetishism is not a good standard for determining what is virtuous.

These criteria can be seen as modern expressions of the traditional view that moral prudence is corruptible and needs to be safeguarded. Thus the Aristotelian tradition of "building a fence around prudence" (similar in some ways to the ancient Jewish tradition of "building a fence around the law") has been extended on the basis of the insights of modern critical consciousness.

The repression, alienation, guilt, and fetishism tests can be of great practical service in the effort to cultivate the life of virtue in the midst of an imperfect world. Because these four tests are based on lessons learned from history and experience (especially the experience of self-deception), they can warn us even when our rational capacities are failing to issue warning signals and when it is still too early to notice the harmful effects of our operative attitudes and actions. At a theoretical level, harmful counterfeits and semblances can be distinguished from genuine virtues in light of an ultimate telos, or derivatively, in light of that which facilitates the achievement of this telos (such as the capacity for prudence). In practice, the achievement of a telos (whether personal, social, or ecological) is often so far away or so complex a matter that it is hard to spot counterfeits and semblances simply on the basis of their immediate results. At the same time, we often cannot afford to "wait and see," for by that time it might be too late. This is especially true in the context of the ecological crisis, which in several areas is heading toward a point of no return. Thus listening to lessons learned from history and experience, lessons that are corroborated by wisdom stored in moral tradition, may in some cases be the only way in which we can recognize moral mistakes in a timely manner.

In the case of the four psychological tests discussed above, this means concretely that in our efforts to cultivate virtue we should be on the alert for signs of repression, alienation, unresolved guilt, and fetishism. The ancient Greeks prescribed temperance as a safeguard for prudence, and ever since Western culture has been on the alert for signs of undue passion; a modern critical consciousness would prescribe self-examination, informed by critical social consciousness and followed by therapeutic as well as social corrective measures.[16]

Notes

1. References in parentheses within the text of this chapter are to Murray Bookchin, *The Ecology of Freedom: The Emergence and Dissolution of Hierarchy* (Palo Alto, CA: Cheshire Books, 1982).
2. I use the term "counterfeit" in a different sense here than Lee Yearley does in his book *Mencius and Aquinas: Theories of Virtue and Conceptions of Courage* (Albany, NY: SUNY Press, 1990). Yearley uses the term to describe apparently virtuous activities intended to deceive others, such as a calculating kind of amiability (19). He also observes that "the concept of counterfeit virtues is easily understood

and reveals little of theoretical interest about virtues" (20). For this reason I prefer to reserve the term "counterfeit" for the more interesting phenomenon of terms and their referents that are mistakenly considered virtuous or vicious.

3. On the notion of semblances, see Yearley, *Mencius and Aquinas*, 19–23. I am much indebted to Yearley for drawing my attention to this neglected aspect of virtue theory.
4. In *Remaking Society* Bookchin writes, for example, "Every vice that, in earlier times, was seen as the apotheosis of evil has been turned into a 'virtue' by capitalist society" (Boston: South End Press, 1990), 46.
5. *The Ecology of Freedom* treats the following terms as counterfeit virtues: reverence/awe, obedience, mastery, charity, and simplicity. Bookchin has no positive use for these terms or their synonyms. Not only does he dislike the terms (as sometimes happens when we have been "preached at" too much), he also dislikes what they express. Thus they are counterfeits and not virtuous in any sense. In addition, sensuousness and permissiveness are treated as counterfeit vices. *The Ecology of Freedom* also identifies semblances of the following virtues (no semblances of vices are mentioned): ecological sensibility, holistic thinking, wisdom, honor, (having) personality, and restraint.
6. See Aristotle, *Nichomachean Ethics*, trans. W. D. Ross (Oxford: Oxford University Press, 1925), 1118^{b}1–5.
7. The psychological dynamics of repression play a key role in Bookchin's social analysis. See, for example, 65, 82, and 117.
8. More implicit in this discussion is also the idea that a genuine virtue will not suppress the operation of our senses.
9. See, for example, James D. Wallace, *Virtues and Vices* (Ithaca, NY: Cornell University Press, 1978), 60.
10. See Aristotle, *Nicomachean Ethics*, 1140^{a}25–30.
11. Of course it is possible to disagree about the exact levels that are cause for alarm. Especially the traditional suspicion of strong sexual desire may seem excessive to many modern persons.
12. In *Toward an Ecological Society* (Montreal: Black Rose Books, 1980), Murray Bookchin seems to treat "voluntary simplicity" as a semblance, however. He rejects the way in which the expression is interpreted in reformist environmentalism, but seems to imply that there might be a legitimate use for the phrase (53).
13. This statement reflects Bookchin's assumption that love presupposes a significant level of differentiation. Alternative conceptions of love may not be as inappropriate in relation to nature.
14. For an alternative discussion of modern interpretations of honor, see Peter Berger, "On the Obsolescence of the Concept of Honor," in *Revisions: Changing Perspectives in Moral Philosophy*, eds. Stanley Hauerwas and Alasdair MacIntyre (Notre Dame, IN: University of Notre Dame Press, 1983), 172–81.
15. Yet the text contains many more counterfeits and semblances from which different criteria can be gleaned.
16. Ironically, if Bookchin's specific judgments about sensuousness are correct, this also means that the ancient prescription of sexual temperance (chastity) as a safeguard for prudence must be regarded with suspicion. To what degree was this prescription the work of an anxious, repressed psyche? What good reasons are there to privilege abstract knowledge over the experiential, even bodily, knowledge that belongs to sexual passion?

Chapter 6 Six

ECOLOGICAL VICE: THE SEVEN DEADLY SINS REVISITED

IN THE PREVIOUS CHAPTER I focused on a question with a very practical basis: How can we distinguish genuine virtue from that which is not virtuous? I affirmed the ancient Aristotelian insight that we can only recognize virtue if our capacity for moral reasoning is intact. This insight carries over into our modern age with added force, since psychology and critical social theory have drawn attention to more and subtler ways in which prudence can be corrupted than Aristotle could ever have fancied.

In the current chapter I focus on a related practical question: How can we distinguish genuine vice from that which is not vicious? We can assume that, like genuine virtue, genuine vice can only be recognized if our capacity for moral reasoning is intact. Another necessary condition, however, is that our moral reasoning be informed by clear criteria. Thus we move from the psychological boundary of the life of virtue to its logical core. Can we identify criteria that define the concept of vice in an ecological age? In this chapter I will survey ecological interpretations of a well-known traditional vice catalogue, the list of Seven Deadly Sins, in order to show that time-tried criteria for the presence of vice, criteria that can inform our moral reasoning, are still operative. Since these criteria are affirmed by the wisdom of a long tradition as well as by the sensibilities of ecologically committed persons, they should be relatively trustworthy. In the meantime we will also discover how ecowriters apply old criteria for distinguishing vice in rather innovative ways. These innovations will help us fine-tune an ecological understanding of the concept of vice.

NEW MEANINGS

The (in)famous list of Seven Deadly Sins has for many centuries guided Christian moral self-examination.[1] As it happens, all the Seven Deadly Sins—

pride, envy, anger, sloth, greed, gluttony, and lust—now reappear in some form or another in the discourse of ecologically minded people. (I will focus here on ecotheologians and ecophilosophers.) As we might expect, however, these traditional vice terms (and their synonyms) no longer mean exactly what they used to mean. Let me therefore begin with a survey of how the Seven Deadly Sins reappear in ecoliterature, after which we can meaningfully inquire whether these modern interpretations still express traditional formal criteria.

The vice of pride or *hubris*, first on the traditional list as the most pernicious of sins, is a major concern in the environmental movement. Christian as well as non-Christian ecowriters warn against the prideful attitude that makes us humans think we are number one in the universe—that we are, in the words of ecofeminist Charlene Spretnak, "central and in control."[2] Ecologically minded people often refer to this form of pride as human arrogance or anthropocentrism.

Such an attitude is harmful, the argument goes, because it ignores the intricate network of relationships in which humans find themselves. As a result, human actions are not geared toward sustaining this network; they do not contribute to what ecological prudence recognizes as a good. In fact, anthropocentrism may lead directly to abuse, for example, when people find their own interests so important that they are prepared to let the dictates of the market economy overrule the limits of ecosystems and the biosphere. Of course in the long run such hubris will backfire, ecowriters warn. Nature itself will "pull down the mighty from their thrones and raise up the humble"—that is, it will undercut those who have sought to lift themselves above the structures of biological life (the Latin term for pride, *superbia*, translates the Greek *huperbios*, which means "above life"), and it will support those who have remained close to the ground (the Latin term *humilitas* literally suggests closeness to "humus," i.e., "soil" or "ground"!).

This argument illustrates how, in terms of meaning, the vice of pride is radically transformed by an ecological perspective. The main concern is no longer that humans may be usurping God's place in a hierarchical order of being, but rather that humans do not keep their proper place within a web of earthly relations.[3] And rather than warning that pride leads to the fall of the individual soul, ecowriters warn that it may cause the fall of creation (though perhaps not of the entire creation, since it would be prideful to claim that humans have such destructive power!).

In contrast to the term "pride," the term "envy" is rarely mentioned in environmental literature. This silence may reflect the general modern hesitancy to see envy as a moral defect. Yet environmentalists do denounce related attitudes, such as competitiveness and resentment about the limitations inherent in the human condition.

Competitiveness is especially a concern when, fueled by little more than envy of another's success, it leads to ecologically harmful behavior. For example, a neighbor striving to match the bright green lawn next door may, without second thought, opt to use the most aggressive herbicides and fertilizers on his or her own lawn. In the business world too, where competitiveness has become a creed that far exceeds the demands of the bottom line, lesser known ecologically responsible options are often ruled out from the start in favor of invasive options with a proven record of reliability. Also, like pride, competitiveness is ultimately self-destructive. Inspired by Gaia theory, Rosemary Ruether points out:

> Any absolutization of competition that causes one side to be wiped out means that the other sides of the relation thereby destroy themselves as well. The falsity of the human cultural concept of "competition" is that it is mutually exclusive. It imagines the other side as an "enemy" to be "annihilated," rather than an essential component of an interrelationship upon which it itself interdepends.[4]

This is not to say that all competition is vicious. Ruether sees a positive role for "interdependent competition" within the context of cooperative relationships. As with many vices, harmfulness lies in extremes.

Resentment about the limitations inherent in the human condition can be equally destructive from an ecological perspective. Through technology, humans try to push these limits ever further, without sufficient attention to the consequences for the Earth community as a whole. As embodied beings, humans are relatively place-bound, yet enviously pursue an ideal of mobility and freedom that is only possible through excessive burning of fossil fuels. As living beings, humans grow older, yet modern Westerners envy the young and create a fiction of youth for themselves by means of an endless line of products. As mortal beings, individual humans will be forgotten after a few generations, yet they envy the famous of the past and engage in actions which, like the carvings in a tree, may be all too lasting in their effects on the environment.

By denouncing competitiveness and resentment about limitations, ecologically minded people clearly sound a countercultural note. Their reflections suggest specific new ways in which the old vice of envy regains moral significance in the midst of an environmental crisis.

The third Deadly Sin, anger, has positive as well as negative moral connotations in ecological discourse. Anger as justified indignation expressed in an appropriate manner is generally considered virtuous, especially if it provides the spark that leads to rectifying action. Dorothy Dinnerstein recommends "using our rage, like our shit, not for ammunition as our monkey cousins do, but humanly: to fertilize green growth."[5] In an ecological age there is much occasion for such anger. For example, it would be entirely appropriate to be angry at the company that has knowingly violated legal and ethical guidelines by dumping toxic waste in an unsafe and unmonitored location. And if this anger leads to a campaign to discourage such violations, all the better.

Yet in an ecological age, certain expressions of anger may also be inappropriate. Environmentalists frequently denounce the aggression with which humans treat nonhuman nature (as well as each other). The images of all-terrain vehicles (ATVs) destroying a dune habitat, or people acting cruelly toward animals, come to mind. While not all aggression is rooted in anger, unresolved anger can express itself as aggression, especially toward what is perceived as an inferior or weaker party. Unfortunately, nonhuman nature is commonly seen in such an inferior position and as a result ends up being the target of much abuse.[6]

The term "sloth," having an old-fashioned ring to it, only rarely appears in environmental literature.[7] Yet two related dispositions, which during the Middle Ages were merged to yield the single vice of sloth, are major concerns, namely, pessimism (*tristitia*) and apathy (*acedia*).[8]

At a time when it is not at all certain whether and how the unprecedented breakdown of planetary systems can be halted, pessimism is an understandable and, in a certain sense, realistic human response. Yet ecowriters tend to disapprove of such an attitude. As J. Baird Callicott puts it, "There's no survival value in pessimism. A desperate optimism is the only attitude that a practical environmental philosopher can assume."[9] One argument against a pessimistic response is that it guarantees that even the few remaining opportunities for reversal will be missed. Christians committed to a vision of a transformed world have another reason to avoid despair. Their faith is, and has always been, a resistance faith against seemingly overwhelming odds. To fall into despair is therefore sinful, both in the sense of being ecologically irresponsible and as the denial of a key Christian vision of divine promise and grace. Thus, for pragmatic as well as theological reasons, sloth in the sense of pessimism remains a moral concern in an ecological age.

Sloth in the sense of apathy also retains its relevance. Apathy may be the consequence of despair, yet it may also stem from other causes such as denial, shallow thinking, smugness, or even plain being too busy and tired to change anything. It is a major problem from an ecological perspective, because turning the crisis around will require extensive social and personal changes. How to get consumers to break their household habits and recycle their wastes, how to get voters to become informed about environmental legislation, how to encourage commuters to walk, bike, or use public transportation rather than take the car—in short, how to mobilize and motivate people to become ecological healers, are questions of foremost importance. In taking up these questions, ecowriters constantly encounter a wall of apathy.

Greed, the fifth Deadly Sin, is also a major concern in ecoliterature. Authors commonly observe that excessive human greed cannot be sustained by the life systems of the world. (In fact, at this point many environmental experts even question whether simple basic needs can be met for all people at current population levels—let alone for other species, many of which have already been extinguished.) Socially aware authors also observe that greed tends to enhance economic and political inequities, which, in turn, can harm the environment. For example, in developing countries, profit-driven agribusinesses may force poor farmers onto hilly terrain, resulting in increased poverty as well as erosion.

Much greed today takes the form of consumerism, aptly described by Murray Bookchin as "consumption for the sake of consumption."[10] As many ecowriters point out, this largely modern phenomenon is fueled by aggressive advertising, yet it may have deeper roots in alienation, insecurity, loss of the ability to provide basic necessities at a local level, as well as larger structural problems.[11] Overall, ecowriters (as well as Pope John Paul II)[12] agree that consumerism harms the environment. It is not difficult to see why this would be the case. As farmer-philosopher-poet Wendell Berry points out, consumerism is highly "unnatural" in the sense that "a consumer is one who uses things up, a concept that is alien to the creation, as are the concepts of waste and disposability."[13] In other words, consumerism harms the environment exactly because it does not "fit in"; it is, to use traditional terms, a violation of a "natural law."

"Gluttony," like the word "sloth," sounds terribly old-fashioned. Not surprisingly, this dated term hardly appears in environmental literature. An interesting exception can be found, though, in Wendell Berry's *A Continuous Harmony*. Berry shows how we can extend the meaning of gluttony to include such things as the frivolous consumption of electricity.[14] Elsewhere, he even suggests that "the desire to get to heaven, if it involves neglect of the life of the world, becomes only a rarefied form of gluttony."[15] Closer to the traditional meaning of the term, though I have not seen it explicitly expressed as such, would be the idea that eating high on the food chain is a form a gluttony. Not all ecowriters are vegetarians, but all agree that the Western habit of eating diets rich in meat, poultry, and large fish is unsustainable and must be changed. This call for change parallels traditional warnings against indulging in too much "luxurious" food.

Still, of course, the underlying motivations are not traditional. For example, over the centuries Christian practices of fasting and general temperance in eating and drinking have been intended first and foremost to serve and preserve the good of the soul. Too much good food was thought to lead to weakness of will, lust, and irrationality, all of which could threaten a person's salvation. By contrast, modern ecowriters warn against eating too high on the food chain primarily because it threatens planetary flourishing, or because it involves killing sentient animals. Also, at a personal level health concerns have taken the place of concern for salvation in the afterlife. Those authors who do draw a link between eating habits and a person's spiritual state of being also always draw the link between eating habits and the state of the planet.

The last vice in the traditional catalogue of Seven Deadly Sins, lust, seems hardly a concern at all in ecoliterature. In fact, one commonly sees lust—or related concepts such as *eros* or sensuousness—celebrated as a virtue. Lust is perceived as a natural human capacity for attraction to another, a capacity that not only comes into play in genital-sexual encounters but in any intimate human engagements with the created world. Since ecowriters tend to value things natural, it stands to reason that they would also value lust as a natural phenomenon. Many Christian as well as non-Christian ecowriters openly celebrate the power of erotic attraction. Thomas Berry, for example, suggests that the "immersion into matter" that characterizes sexual passion might be considered "as immersion into the divine."[16] Other authors propose that erotic power be mobilized in the struggle to build sustainable life communities.[17]

Yet even from an ecological perspective, lust can be problematic. This is particularly true when sexual passion gets out of balance and begins to cloud common sense. Especially at a time when chances for healing the planet, to a large extent, depend on population control, it is easy to see how the heat of passion, if not appropriately channeled, can lead to irresponsibility with not only interpersonal but also global consequences. In such a case, lust takes on an inappropriate dimension. It is important to point out, however, that this note of caution has nothing to do with the view that sexual desires are low, dirty, or animal-like, and hence despicable (a view that has held sway in much of Western tradition). Lust is only problematic insofar as it may obstruct procreative responsibility.

Clearly, from the vantage point of ecological consciousness, the Seven Deadly Sins of pride, envy, anger, sloth, greed, gluttony, and lust no longer mean what they used to mean. Although there may be some basic continuities—for example, pride is still a relational vice, and greed still refers to an excessive desire to amass possessions—changes of meaning are pervasive as well as significant. From an ecological perspective, the content of all the vices is calibrated in relation to earthly or even cosmic flourishing. Compared to a tradition that focuses mainly on interhuman and human-divine relations, this new frame of reference represents an extension[18] that cannot fail to affect every vice at the core of its meaning. Also, some vices, like anger and lust, are overshadowed in significance by the positive counterparts that go by the same name, a shift that again reflects a general change of preoccupations.

If the content of the Seven Deadly Sins has changed so significantly, however, we may well ask whether they can still be called "vices" in the familiar sense of the word. Do these new interpretations still express traditional formal criteria?

Traditional Structure

Within the Aristotelian-Thomistic tradition of virtue ethics it is generally assumed that two criteria must be met in order to speak of genuine vice. First of all, the attitudes and actions of the moral agent must endanger some internal or external good; there must be *harm*. In this respect, discerning genuine vice is a matter of judging results (though it remains a point of debate whether these results can be judged by an outside observer). However, the term "vicious" not only connotes "harmful," but also "malicious" or "mean-spirited," and the tradition recognizes that a bad act alone does not necessarily mark an agent as a vicious person. For one thing, the agent may have made an uncharacteristic slip, most likely under unusual circumstances, as when a generally frugal person goes on a ruinous shopping spree upon hearing that she has been deserted by the love of her life. Or the agent may not have known about the harmful consequences of his attitudes or actions, as when a young husband who had been raised to adopt macho values is surprised to learn that what he saw as disciplining his wife is called spousal abuse. And then, of course, there is the tragic case in which the agent has no other option but to do harm, as when a single parent under pressure to make a living cannot pay enough attention to her children. In none of these cases we would speak of vice, because the agents have not deliberately and consistently rejected good for evil. In other words, to speak of vice there must be *choice*, which is the second formal criterion.[19]

The two traditional criteria of harm and choice have a long history. In Aristotle's *Nicomachean Ethics* we find them at the beginning of his discussion on moral virtue in Books II and III. In Book II Aristotle focuses on the "external" questions of good and harm. Here vice is defined as the harmful extremes of excess and deficiency. In Book III he focuses on the "internal" questions of praise and blame. Here vice (as well as virtue) is described as the result of choice, which he defines as "deliberate desire of things in our own power." Hence the common practice of blaming and punishing the vicious makes sense.[20]

The criteria of harm and choice have also drawn much interest in the recent revival of Aristotelian ethics among philosophers, although they are generally not discussed together as criteria of vice. In fact, noticing the conspicuous absence of an entry for "vice" in the indices of many key works in this literature, I must concur with Lee Yearley's observation that "few philosophers . . . have focused on the analysis of vice. The inattention to vice is surprising, given the generally accepted notion that we can illuminate the conceptual form of perfected states by examining failures in them."[21] However, interesting refinements of the criteria of harm and choice do occur in the literature, although they are usually hidden in discussions of other topics. Some of these refinements highlight basic differences with ecological literature, and some highlight important parallels. Let me give an example of each.

The question of the harm involved in vice is insightfully explored by Nancy Sherman in her book *The Fabric of Character*. According to Sherman, viciousness is "bad" because it impairs coherent character development, our capacity for practical reasoning, and our perception of the human good. But most deplorable is the fact that "for the vicious, the means are not worthwhile without the external end."[22] Sherman's psychologically rich insight helps us appreciate the nuances of the harm involved in vice—that is, the harm done to humans. Although it may be true that those who have become impaired and calculating because of their vicious habits are likely to form a threat to nonhuman nature as well, this is not directly on Sherman's mind (nor, for that matter, on the minds of most other philosophical virtue ethicists). Not surprisingly, we will find a rather different set of concerns when ecowriters take up the question of harm.

The question of the choice involved in vice continues to spur discussions on the topic of freedom and its necessary conditions. Increasingly, philosophers draw on the discourse of clinical and developmental psychology in this context, suggesting an idea that I will call "derivative responsibility": While we may not directly choose to be vicious, insofar as we have chosen the paths that lead there, we are responsible for our vices.

Neil Weiner, for example, argues that "vice, for Aristotle, is some kind of pain relief." We tend to deal with pain, especially the pain we refuse to acknowledge, through compensation. This is how we fall prey to excess, to unharmonious living. Yet Weiner, with many therapists, argues that we are responsible for the repression of our pain, since it is "willed ignorance." In that sense we are, derivatively at least, responsible for our vices.[23] Terence Irwin, looking at the development of character, arrives at a similar conclusion. He writes:

> Aristotle's search for the origin of an action leads him beyond the actual causal sequence to the agent's attitude and character. Even if I act without deliberation and premeditation on a sudden impulse of emotion or appetite, the origin may still be in my character and decision; for the presence or strength of my desire may be the result of the character and

> decisions I have formed. I may have deliberately cultivated this sort of impulse, or I may have failed to do what I could reasonably be expected to do to prevent its growth. If my voluntary actions are related in this way to my decision and character, their origin is in *me* in the relevant sense, and I am fairly held responsible for them.[24]

Thus, following a period in which the therapeutic excuse was fashionable, the idea of derivative responsibility now increasingly occupies the attention of moral philosophers. As we shall see, a similar trend marks ecological literature.

The criteria of harm and choice are also underwritten by the Thomistic side of the tradition; both are typically defined in relation to God. Thomas Aquinas, following Augustine, observes that "two things concur in the nature of sin, viz. the voluntary act, and its inordinateness, which consists in departing from God's law."[25] In the subsequent manualist tradition of Catholic moral theology, the criterion of harm became known by the name of "serious matter," and the criterion of choice by the combination of "sufficient reflection" and "full consent of the will."[26] These criteria are still used in current, more systematic discussions of sin.[27]

Lately the manualist tradition has been widely criticized for encouraging a preoccupation with sin as an objective act of an isolated individual. Alternatively, many Catholic moral theologians and Protestant ethicists redefine sin in terms of a failure to be responsible in relation to God and creation.[28] Such a failure is usually a matter of a pattern of decisions leading to serious character deformation, rather than an isolated act.[29] These criticisms and alternative views do not represent a rejection of the traditional criteria of harm and choice, but rather a reaffirmation that corrects imbalances introduced by the manualist tradition, which did not always hold both conditions together. Richard Gula, for example, stresses that the traditional conditions of sin are "still valuable."[30] They just have to be interpreted in an integrated, person-oriented way. This means less focus on external acts than has been the case in the traditional confessional setting, and more focus on character and motivations.

In the remainder of this section I will use the two traditional criteria of harm and choice as touchstones to examine the formal structure of ecological vice language, which on the face of it looks so untraditional. Specifically, I will pursue the question, What makes ecological vices "morally bad" (i.e., blameworthy and changeable, as opposed to just pitiable and regrettable)? Admittedly, this is a rather limited, mostly conceptual question. Much more could be explored on the topic of vice, including the physical, psychological, and communal conditions of choice; the narrative sources supporting our view of what is harmful; and the character development that leads toward or away from vice. One could even argue that without such elements of "thick description," the formal question of the "morally bad" cannot be meaningfully pursued. As should become clear, however, my query will rely on thick description; the limitation of the question is a matter of focus, not of rejecting the richness of the relevant context.

Do ecological interpretations of the Seven Deadly Sins meet the criterion of harm? From the previous discussion it should be clear that the traditional vice terms on this list (or conceptually related terms) continue to express danger to some good, even though some also carry a positive meaning. Through pride we ignore the network of earthly relationships, thereby risking its destruction. Through envy we engage in competition that spirals beyond and destroys cooperative structures. Envy also entices us to engage in the dangerous game of defying natural limits. Our anger becomes harmful when it is vented on "weaker" humans and nonhuman nature. Through pessimism and apathy we miss vital opportunities for future survival. Greed causes us to bust the great cycles by which nature regenerates itself. Through our gluttony we endanger the world's food supply and cause unnecessary suffering to innumerable animals. And lust helps us to multiply to such an extent that we threaten to suffocate the Earth.

These dangers, of course, are not exactly what worried the medieval interpreters of the catalogue of Seven Deadly Sins. Ecoliterature takes its bearings from the ecological goods of bioregional, biospheric, or even cosmic flourishing. To the extent that some ecowriters see no divine participation in this flourishing, and to the extent that sin could be defined as that which harms God who is the only true good,[31] one could indeed argue that in their writings the Seven Deadly Sins no longer represent genuine sins. Yet insofar as ecological good counts as a genuine good (a good that includes human flourishing and that for some authors also involves a divine dimension), and insofar as the Seven Deadly Sins harm this good, they continue to meet the first criterion of genuine vice.

The second criterion, the criterion of choice, can also be identified in the literature, which generally conveys a strong sense that alternative ways are possible. The vices of greed and sloth provide good examples. Ecowriters may point to the deeper, often pathological and social causes behind consumerism, and they may use terms like "compulsion" and "obsession" to describe the behavior of the avaricious individual,[32] but ultimately their writings do not "explain away" the vicious nature of the phenomenon. Greed may be understandable, but that does not mean that it is inevitable. When greed is bewailed in the literature, the tone is not one of resignation, but one of criticism and exhortation. Ecowriters seem to assume that people always have a "fundamental option" not to buy into the consumerist mentality.[33] This option is illustrated by the witness of those who have successfully chosen a different path of frugality and resourcefulness. Similarly, pessimism may be understandable in an age faced with serious ecological problems, yet the hope-filled activism of ecologically minded people proves that it is not a necessary response. And apathy may be the understandable result of pessimism, denial, shallow thinking, smugness, or preoccupation, yet all of these causes can, in principle and with the right means, be removed—for example, by seeking therapy, or by becoming better educated, or by engaging in meditation, or even by making a start with small constructive actions such as planting a family vegetable garden or starting a neighborhood cooperative.[34] The list goes on, but I think the point has been made: Ecoliterature generally expresses the assumption that there are realistic positive alternatives to our harmful attitudes, which makes our persistence in these attitudes a matter of choice.

We may conclude, then, that in an ecological age the Seven Deadly Sins remain genuine vices in the traditional sense of the word. In addition to the social, psychological, and spiritual harm for which they have traditionally been feared, all seven dispositions can be linked to environmental destruction.[35] They are deadlier than ever.[36] Moreover, all involve attitudes and actions that are avoidable and for which people can, in some meaningful way, be held responsible. They are still "sins."[37]

Innovating Tradition

So far, my line of argument can be expressed by a simple formula: The fate of the Seven Deadly Sins in ecoliterature suggests that ecological vice language is materially innovative, but formally traditional.[38] A few qualifications are in order, however. After all, "traditional" is a loaded term. Also, I do not intend for my argument to be taken as proof of the persisting unique importance of the catalogue of Seven Deadly Sins throughout the permutations of history (I consider other vices such as cruelty, neglect, and betrayal just as important).[39] My aim is not to make a conservative claim or a pious point, but rather to understand what Robert C. Roberts would call the "grammar" of ecological vice language—its underlying patterns and assumptions.[40] This requires that I further explicate the claim that this language is formally traditional.

As I have already indicated, the traditional criterion of choice is in flux these days, with philosophical attention shifting from direct choice to derivative responsibility. The same shift can be observed in ecological literature when authors suggest that measures such as therapy, education, and meditation may be necessary to prevent and overcome ecologically harmful habits. Let me explore this shift in more detail with respect to the sins of pride, greed, and lust.

Ecologically minded people tend to see anthropocentric pride as a symptom of a far-reaching alienation, a lack of rootedness within the Earth community. This harmful attitude is vicious, not in the sense that it presupposes full moral culpability, but mainly in the more restricted sense that anyone who becomes aware of the underlying alienation can be held morally accountable if they do not seek ways to overcome it. In other words, humans may not be directly responsible for their pride, but they are responsible *for eliminating to the best of their ability the conditions of alienation that give rise to this harmful attitude.*

Similarly, exercising the fundamental option to refuse consumerism may require a deeper personal transformation than a "just say no" approach. This is no ordinary responsibility, but an ultimate life choice, which can be likened to the choice of a prison convict to stop participating in the cycle of crime. Much therapy and many setbacks may be required to create the possibility of resisting the perpetual lure of the shopping mall. But where this possibility exists, we can speak of responsibility. And those who act on it can be commended for a truly remarkable effort.

Derivative responsibility may extend beyond the personal realm into the communal sphere. The sin of lust provides a case in point. Many people today do not have the option of family planning due to lack of education, insufficient economic security, or violations of women's rights. Such people should not be labeled as "lustful" if they have large families. Yet insofar as there are realistic social alternatives that would empower these people in the area of reproduction, we can still speak of *collective* responsibility, a form of derivative responsibility (and the basis of the Christian notion of "social sin").[41] Thus lust, which used to be the most private of vices, in an ecological age takes on a public dimension. Societies and institutions that ignore and prevent the conditions necessary for reproductive responsibility commit the ecologically deadly sin of lust.

In sum, if we can extrapolate on the basis of the Seven Deadly Sins, then ecological vice language depends significantly on the idea of derivative responsibility, whether personal or collective. The saying, "tout comprendre, c'est tout pardonner" (to understand everything is to forgive everything), which applies to such a large extent to modern therapeutic culture, does not seem to apply to ecoliterature (which, in fact, puts it at risk of sounding "preachy"). Instead, ecowriters take the expression "you are responsible for your acts" to its furthest level of meaning when they suggest that people are even responsible for those harmful effects of their acts (or lack of action) which they did not directly intend but which they could have foreseen and prevented.

Such a view creates a moral obligation to be informed about risks and to take general preventive measures, such as engaging in systematic (self) reflection, seeking counseling, and other forms of therapy. Thus ecological vice language reflects a move away from the traditional "principle of double effect," which allows foreknowledge without moral culpability. At the same time it reflects a move toward the Western model of understanding the responsibility of corporate executives and public officials, who are often held accountable for developments they did not intend but could have foreseen and potentially prevented through different policy decisions. With these moves, ecowriters participate in a wider adjustment within the Aristotelian-Thomistic tradition as well. Hence, with respect to the criterion of choice, ecological vice language is formally traditional in the best sense of the word: It is part of a "handing on," a reshaping of the past for the future.

With respect to the second formal criterion of vice, the criterion of harm, ecowriters stretch the meaning of "tradition" to its very limits—and here they have far less company from virtue ethicists. In ecoliterature, the old vice labels appear on actions and attitudes that throughout Western history would have been deemed (by those whose deeming counted) outside the realm of the harmful. As far as Christians are concerned, real harm was harm done to those created in God's image—humans. Unnecessarily damaging plants or animals was often treated as insignificant. At best it counted as an offense to their Creator or to their legitimate human caretakers and users (more systemic ecological damage was not even part of the conceptual framework). Humanists, following classical models, left God out of the picture, but retained the human as the main measure of harm. And still today the human person occupies center stage in the virtue theories of many philosophers and theologians, even if those theories are innovative in other respects. This long-standing pattern leads to a fundamental question: Is what ecowriters label as harmful indeed the kind of harm that counts when we consider the formal criteria for vice?

Here we stand at a juncture in tradition. A positive answer significantly changes the way in which the criterion of harm has been applied for many centuries. This is not an easy change, for it requires a careful recalibration of the notion of harm. Such theoretical work (not without an intuitive basis, though) is currently being carried out by many ecothinkers, and it inevitably brings with it new knotty questions regarding the significance and extent of sentience, who or what can be said to have interests, and the meaning of intrinsic worth and value. Yet even if an ecological concept of vice hinges on answers to these questions, ecowriters themselves are already using vice language that assumes an extended notion of harm. Whether their adoption of traditional terms such as pride and greed can be called formally traditional depends ultimately on our view of the renewability of tradition.

CONCLUSION

In an ecological age, how can we distinguish genuine vice from that which is not vicious? At first sight, it would seem that we have reason to be confused. Ecological literature displays a veritable revaluation of values. Even old vices like the Seven Deadly Sins no longer mean what they used to mean. So how do we know when we are dealing with real vice, when we have to watch out, to mend our ways, to work on our characters? Luckily, old criteria still hold. Vice implies harm. And it implies choice. But here again we find ourselves on shifting ground. Choice is not just a simple matter of choosing to do this or that. Apparently we are accountable for our pathologies, for our ignorance in the face of mounting information, even for our characters. And harm is not cut-and-dried either. Is the imminent extinction of a red frog in the Amazon rain forests, spurred on by our love for hardwood furniture, morally relevant? Is cutting down a row of red corral trees to replace them with purple Jacaranda, simply for the sake of beautifying a college campus, morally relevant?

If we want an ecological virtue ethic that is consistent with the inclusiveness and the this-worldly orientation of an ecological worldview, I think we would do well to go with the flow and face these challenges. Ultimately this means that we have to face ourselves. Tradition may still be of some help in discerning genuine vice. But in order to see really clearly we will have to face our own tricks for avoiding responsibility, and for avoiding the signs of harm all around us.

NOTES

1. For recent attempts to adapt the traditional catalogue to modern developments and needs, see William F. May, *A Catalogue of Sins: A Contemporary Examination of Christian Conscience* (New York: Holt, Rinehart and Winston, 1967); Karl Menninger, *Whatever Became of Sin?* (New York: Hawthorn Books, 1973), 133–72; Henry Fairly, *The Seven Deadly Sins Today* (Washington, DC: New Republic Books, 1978); Kenneth Slack, *The Seven Deadly Sins: A Contemporary View* (London: SCM Press, 1985); Donald Capps, *Deadly Sins and Saving Virtues* (Philadelphia, PA: Fortress Press, 1989); Solomon Schimmel, *The Seven Deadly Sins: Jewish, Christian, and Classical Reflections on Human Nature* (New York: Free Press, 1992). For an extensive historical survey of the seven deadly sins, see Morton W. Bloomfield, *The Seven Deadly Sins: An Introduction to the History of a Religious Concept, with Special Reference to Medieval English Literature* (East Lansing, MI: Michigan State College Press, 1952).
2. Charlene Spretnak, *The Spiritual Dimension of Green Politics* (Santa Fe, NM: Bear & Co., 1986), 27.
3. Some Christian environmentalists would argue that this network itself manifests dimensions of the divine. In this sense then, pride would still signify a skewed relationship with the divine.

4. Rosemary Radford Ruether, *Gaia and God: An Ecofeminist Theology of Earth Healing* (San Francisco: HarperCollins, 1992), 56.
5. Dorothy Dinnerstein, "Survival on Earth: The Meaning of Feminism," in *Healing the Wounds: The Promise of Ecofeminism*, ed. Judith Plant (Philadelphia, PA: New Society Publishers, 1989), 196.
6. Compared to traditional interpretations, ecological interpretations of anger actually seem to increase the significance of this attitude, both in the positive and the negative sense. While Western moral traditions have long recognized the distinction between appropriate and vicious anger (one finds it, for example, in Thomas Aquinas's *Summa Theologiae* II-II.158.2 [trans. the Fathers of the English Dominican Province (New York: Benziger Brothers, 1948)]), a moral consciousness shaped by a strong appreciation for the value of nonhuman nature will discern many more occasions for justified anger, as well as many more harmful expressions of anger, than would have been identified by traditions that have given little moral recognition to nonhuman nature. In other words, when ecowriters include nonhuman nature in the network of morally relevant relationships, they thereby *extend the range* of anger, both as a helpful and as a harmful disposition.
7. Dorothy Dinnerstein, for example, refers to "mental sloth," which, along with cowardice, she characterizes as a "psychopathology" ("Survival on Earth," 196).
8. The merger may have been motivated by a desire to reduce the list of eight vices to seven, so that each day of the week could be dedicated to one vice in the interest of a systematic examination of conscience (See Capps, *Deadly Sins and Saving Virtues*, 12).
9. J. Baird Callicott, "The Land Ethic Today," *Topoi* 12 (1993), 50. Process theologian John Cobb calls despair in the face of the environmental crisis "worse than useless" (*Sustainability: Economics, Ecology, and Justice* [Maryknoll, NY: Orbis Books, 1992], 18).
10. Murray Bookchin, *The Ecology of Freedom: The Emergence and Dissolution of Hierarchy* (Palo Alto, CA: Cheshire Books, 1982), 68.
11. See, for example, Bill Devall and George Sessions, *Deep Ecology* (Salt Lake City, UT: Gibbs Smith, 1985), 68. Ironically, consumerism may also be enhanced by the recent phenomenon of environmental marketing. See L. M. Benton, "Selling the Natural or Selling Out? Exploring Environmental Merchandising," *Environmental Ethics* 17 (1995), 15.
12. John Paul II, "The Ecological Crisis: A Common Responsibility," World Day of Peace Message, 1 January 1990 (USCC Publication No. 332–9), 11.
13. Wendell Berry, *A Continuous Harmony: Essays Cultural and Agricultural* (San Diego: Harcourt Brace Jovanovich, 1972), 111.
14. Ibid., 181.
15. Ibid., 9.
16. Thomas Berry, *The Dream of the Earth* (San Francisco: Sierra Club Books, 1988), 151.
17. See, for example, Bookchin, *The Ecology of Freedom*, 325ff. Bookchin here draws on precedent set by a long tradition of utopianism in which sensuousness played a significant role.

    One also finds an appreciation for the transformative power of eroticism in deep ecology literature, which otherwise differs significantly from Murray Bookchin's social ecology. For example, Bill Devall explains that deep ecology warriors "have affective, erotic relationships with bioregion, river, mountain, and they are passionate, aroused—sometimes even angry" (*Simple in Means,*

Rich in Ends: Practicing Deep Ecology [Salt Lake City, UT: Gibbs Smith, 1988], 200–1).

18. From a traditional Christian point of view the change may represent a reduction, however, insofar as not all ecologically minded people include God in the network of earthly or cosmic relations.
19. For Paul Ricœur, the element of choice in sin is symbolized by the "Adamic myth." See Paul Ricœur, *The Symbolism of Evil*, trans. E. Buchanan (New York: Harper & Row, 1967), pt. 2.
20. Aristotle, *Nicomachean Ethics* (trans. W. D. Ross [Oxford: Oxford University Press, 1925], II.2, III.3, III.5. Interestingly, Aristotle understands "deliberate desire" negatively, as desire exercised without compulsion or excusable ignorance. Hence he does not have to commit himself to a strong notion of "free choice."
21. Lee Yearley, "Recent Work on Virtue," *Religious Studies Review* 16, no. 1 (January 1990), 3.
22. Nancy Sherman, *The Fabric of Character: Aristotle's Theory of Virtue* (Oxford: Clarendon Press, 1989), 108–17; citation from 116. See also: "Vicious actions do not seem to constitute their own ends. . . . Their point and focus seem to remain on the external end—power, gain, deceit, avoidance of risk or fear. . . . It is hard to conceive that such individuals would find value in their pursuits if their machinations did not ultimately yield the external ends sought. In contrast . . . the virtuous would still feel their lives had a point and would find intrinsic pleasure in their actions" (115–6).
23. Neil O. Weiner, *The Harmony of the Soul: Mental Health and Moral Virtue Reconsidered* (Albany, NY: SUNY Press, 1993), 138, 147–8.
24. Terence Irwin, *Aristotle's First Principles* (Oxford: Clarendon Press, 1988), 344. Elsewhere Irwin refines Aristotle's notion of responsibility (being worthy of praise or blame) by suggesting that it implies an agent's capacity for deliberation (T. H. Irwin, "Reason and Responsibility in Aristotle," in *Essays on Aristotle's Ethics*, ed. Amélie Oksenberg Rorty [Berkeley: University of California Press, 1980], 117–55). This would exclude young children and animals from the company of the vicious and the virtuous, a notion not acceptable to some ecowriters.
25. Aquinas, *Summa Theologiae* I–II.72.1.
26. See Richard Gula, *Reason Informed by Faith: Foundations of Catholic Morality* (New York: Paulist Press, 1989), 109. Manualist moral theory can also be described as having an objective and a subjective pole, which correspond with the criteria of harm and choice respectively. The objective pole centers around the moral evil of the act itself. This is typically evaluated in terms of its object, end, and circumstances. The subjective pole centers around the evil of the sinner. This is typically evaluated in terms of the will, and to some degree in terms of the intellect (See John A. Gallagher, *Time Past, Time Future: An Historical Study of Catholic Moral Theology* [New York: Paulist Press, 1990], 79, 93).
27. See, for example, Germain Grisez and Russell Shaw, *Fulfillment in Christ: A Summary of Christian Moral Principles* (Notre Dame, IN: University of Notre Dame Press, 1991), 151.
28. Edward Farley even treats "theocentrism" as a distinct *third* criterion for determining sin, in addition to "the differentiation of sin from suffering and tragic finitude" (cf. the criterion of choice), and "the view of sin as distortion of (human) reality" (cf. the criterion of harm) (*Good and Evil: Interpreting a Human Condition* [Minneapolis: Fortress Press, 1990], 125–7).

29. On the Roman Catholic side, the work of Bernard Häring has been especially important in this regard (See Bernard Häring, *The Law of Christ* [Westminster: Newman Press, 1961, 1963, 1966]; and *Free and Faithful in Christ*, vol. 1: *General Moral Theology* [New York: Seabury Press, 1978, 1979, 1981]). The trend to see sin in terms of responsibility has been inspired by the work of the Protestant theologian H. Richard Niebuhr, particularly by his monograph *The Responsible Self* (New York: Harper & Row, 1963). Current Protestant ethics is also profoundly influenced by Niebuhr on this point.

    Protestant narrative ethicists such as Stanley Hauerwas and James McClendon criticize the long tradition of treating vices as the isolated acts of isolated individuals by pointing out that genuine freedom is only made possible by character formation in the context of communal life, and that our understanding of what is good or harmful derives from the narratives that give shape to our communities (see Stanley Hauerwas, *A Community of Character: Toward a Constructive Christian Social Ethic* [Notre Dame, IN: University of Notre Dame Press, 1981]; James McClendon, Jr., *Ethics: Systematic Theology* vol. 1 [Nashville: Abingdon Press, 1986]).
30. Gula, *Reason Informed by Faith*, 109.
31. See, for example, May, *A Catalogue of Sins*, 7.
32. See, for example, Sean McDonagh, *To Care for the Earth: A Call to a New Theology* (Santa Fe, NM: Bear & Co., 1986), 133; T. Berry, *The Dream of the Earth*, 115.
33. Thomas Berry forms an exception in this respect (see chapter 4). Those who are entranced by the myth of Wonderland do not seem to be morally culpable; they do not have a "fundamental option" until they reach a point of crisis. After that, they can be held morally accountable.
34. Arne Naess warns, however, against a traditional method of dealing with apathy that would be self-defeating in an ecological age: "To counter apathy and low self-esteem, moral philosophers have occasionally made the mistake of placing humankind in a unique position in respects incompatible with an open attitude to the rest of nature" (*Ecology, Community, and Lifestyle: Outline of an Ecosophy*, trans. David Rothenberg [Cambridge: Cambridge University Press, 1989], 192). From this sentence one could deduce the principle that efforts to eliminate a vice should not undermine efforts to cultivate an ecological virtue.
35. Some environmentalists see a relation between ecological and spiritual death. Thomas Berry writes, for example: "The trees of the Appalachian Mountains will not be healthy if the rain is acid. Nor will the soil be fertile. Nor will humans have their proper nourishment. Nor will imagination be activated to its grand poetic visions. Nor will our sense of the divine be so exalted if the earth is diminished in its glory" (*The Dream of the Earth*, 79).
36. By referring to the "deadliness" of these sins I also mean to include the traditional meaning of the term "capital vice," namely, a vice that initiates a destructive process by giving rise to many other vices. This remains true for the Seven Deadly Sins (e.g., pride may give rise to exploitativeness, sloth to neglect, greed to carelessness, anger to contempt, etc.).
37. The two traditional criteria outlined in this chapter can also be used in refining the process of identifying new vices, many of which have unusual names or have not yet been named at all. Such a procedure would seem justified in light of my thesis that formal criteria are still operative in ecological vice language at large.

38. Even authors who would on the basis of the content of their worldview be labeled "radical," tend to use traditional criteria to determine the presence or absence of genuine vice. For example, deep ecologist Arne Naess writes: "The process of identification leads us to see much cruelty in nature. But it does not necessarily lead to conceiving any animal to be cruel. A behaviour may be deplored without applying negative ethical standards to the actor" (*Ecology, Community and Lifestyle*, 181). In this example the criterion of harm is satisfied, but the criterion of choice is not, and hence animals cannot be considered "vicious."
39. Judith Shklar makes a case for "putting cruelty first" (*Ordinary Vices* [Cambridge, MA: Belknap Press of Harvard University Press, 1984], chap. 1). I agree with her on the importance of cruelty, but I disagree with her dismissal of pride as only a theological sin against God, and hence not relevant to those who reject "revealed religion." Many theologians in the past have dwelled on the earthly damage done by pride, and now ecoliterature illustrates that pride can even be interpreted as a purely intercreaturely vice.
40. Robert C. Roberts, "Therapies and the Grammar of Virtue," in *The Grammar of the Heart*, ed. Richard H. Bell (San Francisco: Harper & Row, 1988), 149–70.
41. For an analysis of the notion of social sin, see Kenneth R. Himes, "Social Sin and the Role of the Individual," *The Annual of the Society of Christian Ethics* (1986), 183–218.

Chapter 7 Seven

THE GENDER CONSTRUCTION OF VIRTUE DISCOURSE: LESSONS FROM EXPERIENCE

IN THE PREVIOUS CHAPTERS I have shown how tradition can function as a source for critical reflection on ecological virtue discourse. Another source that can provide a whole new range of critical insights is experience. This is particularly true for human experiences of oppression, because these stand as a record of concrete and easily recognizable instances in which the good (of human flourishing in this case) has been violated. In this chapter I therefore pursue the practical questions: What do the hard lessons of experience teach us as we seek to cultivate ecological virtues, and how can we avoid some of the mistakes from the past? My focus is on the oppression of women, whereby I want to heed the ecofeminist suggestion that, if we are serious about our efforts to end the oppression of nature, we should pay attention to what has happened to women in our patriarchal societies.[1]

Unfortunately, virtue discourse has a less attractive side. Building on recent feminist writings, I will show how for centuries it has subtly contributed to the oppression of women in the Western world. If we follow an ecofeminist line of thought, this means that when it comes to the task of creating a truly liberating ecological ethic, virtue discourse is suspect. Yet can it be retrieved? Much depends, I believe, on how we deal with what I call the gender construction of virtue language. If ecological virtues are to be part of a widespread, public ethic, then we must first face the experiences of women within this heritage of gender construction.

GENDER-CONSTRUCTED VIRTUE LANGUAGE

Gender-constructed virtue language divides the life of virtue into four categories, rather than simply the two categories of virtue and vice. There are virtues for women and virtues for men, as well as "typically feminine" vices and "typically masculine" vices. Gender-constructed virtue language occurs in many languages and contexts. Yet it is not always immediately obvious whether four categories rather than two determine the moral life of a particular time or people, because hypocrisy commonly accompanies the multiplication of categories. For example, a virtue may in theory be claimed to apply to all, whereas in practice one sex may be held exempt from the risks and burdens of its cultivation (e.g., sexual fidelity). Thus the identification of gender-constructed virtue language often requires detective work.

Several feminist scholars have engaged in such critical work, motivated by the suspicion that we are dealing with an ideology that has served to oppress women and that needs to be unmasked.[2] Mary Daly voices the suspicion as follows:

> Much of the traditional theory of Christian virtue appears to be the product of reactions on the part of men—perhaps guilty reactions—to the behavioral excesses of the stereotypic male. There has been theoretical emphasis upon charity, meekness, obedience, humility, self-abnegation, sacrifice, service. Part of the problem with this moral ideology is that it became generally accepted not by men but by women, who have hardly been helped by an ethic which reinforced their abject situation. This emphasis upon passive virtues, of course, has not challenged exploitativeness but supported it.[3]

As a result of feminist detective work, it is now possible to sketch the historical development of gender-constructed virtue language in the West. It is beyond the scope of this chapter to provide a full account of this development, but a few highlights will help provide a context for the ensuing argument.

The origin of gender-constructed virtue language must be sought in the belief, held throughout much of Western history, that women are not particularly capable of developing those virtues that are most valued (by the men who do the valuing). Hence the "need" for a separate moral standard for women—a standard commonly tied to their prescribed social functions. As Susan Moller Okin points out in her book, *Women in Western Political Thought*, pre-Socratic Greeks associated the concept of *arete* (excellence or virtue) with the kind of life that was only possible for elite men. Consequently, "'woman's *arete*' was a relative term, consisting of a set of qualities entirely different from those expected of men, who alone could achieve absolute excellence."[4] Following a similar assumption, Aristotle subsequently argued that women could not develop the virtues of justice and friendship to the extent that men could, due to their inferior position in relation to men.[5] Aristotle also associated lack of perseverance, and especially the cowardly act of seeking death to escape from pain, with *malakía* (softness, effeminacy).[6] Women, then, were not only incapable of exhibiting virtuous perfection, but they were also more prone than men to exhibit certain vices.

Christian virtue language has also been marked by gender constructions. The influential virtue theory of Thomas Aquinas provides a good example. As the great synthesizer of Christian and pagan traditions, Aquinas borrows from and expands on the gender-constructed virtue language belonging to these traditions. He approvingly cites Augustine's assumption that "thou shouldst excel thy wife in virtue"[7] and, like Gregory the Great before him, uses the image of "daughters" to discuss the consequences of the capital vices. Aquinas also shares Aristotle's assumption that women are more prone to cowardice and lack of perseverance. To this he adds the opinions of other classical philosophers, who had argued that the Aristotelian virtues of confidence and magnificence are particularly "manly" virtues.[8] Aquinas goes to some length explaining the gender connections in even greater detail than his predecessors had done.

With the political philosophy of Machiavelli, the gender construction of virtue language takes a subtle turn. The idea of moral virtue becomes associated primarily with the "gentler" Christian virtues, which conveniently get relegated to the private realm of the home—the realm of women—where they will not interfere with the power politics of the public world, the realm of men.[9] Morality becomes synonymous with chastity, obedience, self-sacrifice, patience, meekness, and charity: qualities increasingly perceived as "feminine"—that is, as natural characteristics of the ideal image of middle-class and aristocratic women. Courage, magnificence, and other traditionally "masculine" virtues continue to be valued in the public realm, but are no longer considered marks of *moral* virtue—rather, they become simply marks of virility,[10] along with other more questionable traits such as aggression, toughness, domination, ambition, cynicism, and emotional control.

From the time of Machiavelli to the present, much lip service has been paid to the "feminine" virtues of the private realm. Yet in the overall scheme of values, they are actually held in low esteem. Riane Eisler's claim that "'soft' or 'feminine'" values get devalued when masculinity is associated with domination and conquest may offer an explanation of this curious cultural lapse of integrity.[11] At any rate, the association of "virtue" with femininity has not, as might be expected, improved the status of women in Western society. On the contrary, this Machiavellian development in the gender construction of virtue language has contributed to the isolation of women from social structures of power.

Moreover, the Machiavellian association of women with private virtue has resulted in women's alienation from the power of their own embodiment. As Catharina Halkes has observed, "At first, women were perceived as more animal and corporeal, features that simultaneously scared and fascinated men. Yet in the 18th and 19th centuries a change occurs: women are now seen as mistresses of home and family, and they are released from their sexuality."[12] If Halkes is correct, we see that women are increasingly held up as models of virtue, but the ideal of femininity has been purged of sexual connotations. This purging was made necessary by the ancient, but still operative, belief that virtue and sexual passion are mutually exclusive.[13] Thus only when relieved of the curse of Eve and remade in the image of the Virgin Mary do women gain the privilege of being closer to perfection than men. Yet the privilege comes at the price of having to renounce, deny, and suppress embodiment and sexuality as sources of identity, community, and power.

Finally, the Machiavellian association of women with private virtue has also contributed to class and race segregation between women, leaving some women in a doubly or even triply disadvantaged position. As Catharina Halkes continues to observe, the women who during the eighteenth and nineteenth centuries were praised for their domestic qualities belonged either to the middle class or to the aristocracy. Working-class women—and, I would add, women of color—were still viewed with contempt.[14] The portrayal of these women as never within reach of perfection continued more ancient traditions of gender-constructed virtue language. Yet even in ancient times, the imperfect female had had a place in the social order and hence a degree of value. The working-class woman and the woman of color, however, did not fit in the newly conceived private realm. Without spatial legitimation, she had no positive value. Only the so-called "feminine vices" defined her being: If she was dark, or worked in a sweatshop, she was presumably also loose, lazy, or even a lunatic. Thus, the gender construction of virtue language resulted in the segregation of women into two groups, split along class and racial lines: virtuous, "high" women and vicious, "low" women.

This, in a nutshell, is the legacy of gender-constructed virtue language. Rather than improve the life of virtue, as it was ostensibly designed to do, it has contributed to many problems, including the segmentation and privatization of ethics, the hypocrisy of moral ambivalence, the disembodiment of ethics, and moral elitism. All of these have been factors in the oppression of women. In the remainder of this chapter I will show why the heritage of gender-constructed virtue language gives reason for caution as newly emerging ecological virtue language becomes part of a wider public discourse.

"Feminine" Ecological Virtues

When we examine ecological virtue language in light of the Western heritage of gender-constructed virtue language, one thing immediately stands out: Many core ecological virtues would traditionally have been considered "feminine" virtues. For example, ecowriters widely and consistently advocate love, caring, compassion, gentleness, harmonizing, humility, healing, intuitiveness, listening, modesty, nurturing, sensitivity, tenderness, openness, adaptability, and cooperation.[15] Especially since Machiavelli, these virtues have primarily been associated with women.

This historically embedded association invites critical questions. As ecological virtue language gains a wider popular following, will it be construed according to old patriarchal patterns? How can ecologically minded people prevent a repeat of the harms that have accompanied gender construction, a repeat that might extend to the nonhuman world as well? Ecoliterature currently displays four different approaches in dealing with the "feminine" image of key ecological virtues. I will label them the "Go-on Option," the "Head-on Option," the "Dream-on Option," and the "Hands-on Option."[16] As my titles already suggest, I see only the last one as viable in preventing a repeat of patriarchal problems.

One approach is to accept the assumption that certain traits are "feminine," but to deny that these are genuine virtues. In other words, the "Go-on Option" continues the patriarchal tradition of holding that "feminine" traits do not lead to flourishing, or perhaps even form an impediment. According to Josephine Donovan, this is the position taken by animal rights activists Peter Singer and Tom Regan. Singer, she writes, "fears that to associate the animal rights cause with 'womanish' sentiment is to trivialize it."[17] And in *The Case for Animal Rights*, Regan suggests that "'since all who work on behalf of the interests of animals are . . . familiar with the tired charge of being "irrational," "sentimental," "emotional," or worse, we can give the lie to these accusations only by making a concerted effort not to indulge our emotions or parade our sentiments. And that requires making a sustained commitment to rational inquiry.'"[18] Regan's words suggest that a "feminine" image must be avoided, and this can be done by building an ecological virtue ethic around "masculine" traits, such as rationality.

This approach has various problems. First of all, the feminine image of certain traits is rather uncritically accepted. One misses an analysis of how this image got constructed in patriarchal societies, and especially how it acquired such negative connotations. After all, it is not self-evident why being "emotional" should be a liability in social interactions between persons who all have emotions. Second, even from a "masculine" perspective not all traditionally "feminine" virtues can be so easily dismissed. Although it may be possible and sometimes strategically desirable in a patriarchal context to control one's emotions and not come across as "sensitive," how could it ever be advisable to suppress one's ability to feel compassion, to care, to listen? Or worse, who, in the name of ethics, would ever want to pride themselves on not having such abilities at all?[19] An ethic without these virtues would be a heartless monstrosity, something we would have reason to fear (and I do not think Singer and Regan have this in mind). Finally, the "masculine" traits on whose shoulders the entire weight of ecological virtue discourse would come to rest may turn out to be not reliable enough for the task at hand. Control and rationalism—these very traits have been widely implicated in creating our problematic relationships with the nonhuman world.

A second approach is to face the history of gender construction head-on and to affirm that virtues such as caring, healing, sensitivity, and cooperation are indeed "feminine," even quintessentially so, but then to draw the opposite conclusion: that these virtues are key assets in the effort to build better ecological relationships. Central to this "Head-on Option" is the idea that women and nature are integrated in many ways, including through physical cycles, production, and reproduction. Because of this close association, women know how to sustain a flourishing network of human and nonhuman relationships.[20] Many radical and spiritual feminists, including Susan Griffin, Charlene Spretnak, and Vandana Shiva follow this path.[21]

There are several problems with this approach also. Many ecofeminists have objected to the essentialism inherent in the radical feminist position. How do we know that traditionally "feminine" virtues are indeed "natural" for women, and not rather the result of socialization?[22] Being the mother of both a son and a daughter, I personally have become less skeptical about essentialism than I used to be—at least I cannot attribute all the traditional gender differences I see in my children to "nurture," nor to the uniqueness of their characters. Yet I do pose questions at the combination of essentialist reasoning with hierarchical reasoning when specifically "feminine" virtues are seen as superior to all other traits. I agree with Lori Gruen that "by establishing superiority in theory, the groundwork is laid for oppression of the inferior in practice."[23] More specifically, if essentialism implies that only women can adequately cultivate the key ecological virtues,[24] then we would be faced with reverse discrimination in the environmental movement, since men could by virtue of their gender only be second-class participants. Even though women who are currently sidestreamed in the movement[25] may, with justified anger, wish for such a reversal, the implied exclusivism would ultimately undercut our efforts for more inclusive relations with the nonhuman world.[26] Finally, the "Head-on Option" is also likely to have counterproductive results in circles beyond the core of the environmental movement. In the hands of the public media, a strong praise of green femininity will be turned into a praise of pink folly, matched by a reactionary affirmation of blue masculine values. Of course we may choose not to care; but then we have also written off prospects for a wider socioecological transformation.[27]

A third approach involves striving to change the image of feminine ecological virtues so that they can be incorporated into a "gender-blind" ethic that could be followed equally by everyone. This option depends on a commonplace line of "proof": that men can be caring and sensitive, just as women can be rational and courageous. In other words, there are no substantive moral differences between the sexes, and in order to avoid past discrimination we must make our moral discourse as gender-neutral as possible. If necessary, this may involve redefining and renaming some traditionally "feminine" virtues. For example, we may have to clarify that cooperation has nothing to do with "selling out," but that it is rather a form of solidarity. With the exception of some ecofeminists, most ecowriters who value traditionally "feminine" ecological virtues already treat them as gender-neutral. Thus, de facto, this option reflects a majority position, even though few authors would see it as a choice they consciously made.

The approach, though popular, is not quite problem-free either. As Janis Birkeland observes, "A gender-blind prism hides problems centering on power, dominance, and masculinity, and consequently backgrounds certain realities with an impact upon the environment."[28] To be "gender-blind" is nowadays politically correct, just like it is to be "color-blind." It is part of a well-intentioned dream of inclusivity. But in our blindness we ignore the persistence of old injustices, as well as the existence of real differences.[29] That is why I call this the "Dream-on Option." Moreover, if we simply gloss over the long history of gender construction by systematically silencing the adjective "feminine" and fabricating redefinitions of "womanish" virtues, we are not solving problems but rather creating new taboos. And taboos tend to create their own spirals of denial and suppression. This surely cannot be a liberatory route.

A fourth approach involves critically dealing with the historically grown "feminine" image of key ecological virtues. I call this the "Hands-on Option," an option already practiced by many ecofeminists. It implies a permanent awareness of the heritage of gender construction and its victims, lest the old harmful patterns repeat themselves.[30] It implies a relentless effort to change the patriarchal institutions that support harmful gender construction. It implies an honest appreciation of the value of traditionally "feminine" virtues in an ecological age, so we can build wholesome relational networks. It implies that the question of a genetic basis for these virtues in women and men be left open for inquiry rather than dogmatically decided, so we can steer between the extremes of biological reductionism and genetic denial. It implies that we move beyond limiting conceptions of femininity and masculinity, including the related concepts of private and public,[31] so we can avoid the harmful habits of stereotyping and dualistic thinking.[32] And this does not imply gender blindness.

Of course no option is perfect, and I must admit that the "Hands-on Option" suffers from involving too much work. But at least we will know that we are not just chasing a fantasy through smiles and denials, as in the "Dream-on Option." Nor will we suffer from lack of guidance in going about the task of building a gender-conscious ecological virtue ethic. By paying close attention to women's experiences of oppression within a long heritage of gender-constructed virtue language, we can at least learn to avoid the mistakes of the past. In the remainder of this chapter I will infer five boundary conditions from these experiences, criteria that can be used to guide the retrieval of traditionally feminine virtues for an ecological ethic.

Lessons from Experience

(1) *The Inequity Test.* In patriarchal societies, most of the burden of cultivating the "feminine" virtues that are now being reclaimed for an ecological age has fallen on women, exactly because of the perception that this is "women's territory" (nurturing, caring, supporting, etc.). This role diversification has often led to inequity, since women have been expected to do many other work tasks as well. What lessons can we learn here?

First of all, obviously, we should make sure that the cultivation of "feminine" ecological virtues is fairly distributed between women and men. For example, we should not accept any situation in which women do most of the "support" work—that is, as Linda Vance points out, the "actual" work—in the environmental movement while men "are disproportionately valued as spokespersons, theorists, and leaders."[33] If the hierarchical mind-set that puts male above female has been an enabling factor in the exploitation of nonhuman nature, then such exploitation is unlikely to stop as long as, de facto, the mind-set remains alive.

Second, we should be on the alert for parallel inequities between private and public efforts to work for environmental change. As long as "feminine" virtues are perceived as "private," we can expect popular concern for the environment to translate primarily into efforts to change the consumption patterns of the home.[34] But we should not be satisfied with high-visibility recycling and energy-saving campaigns aimed at consumers and schoolchildren, while production and distribution patterns are only marginally altered.[35] Such an imbalance obviously shortchanges the Earth and its life-forms.[36]

(2) *The Trade-off Test.* In patriarchal societies women are often expected to cultivate "feminine" virtues in order to compensate for acknowledged "masculine" vices. Catharina Halkes has, for example, highlighted the traditional assumption that women will caringly keep hot-tempered and beleaguered men on the right moral path.[37] The idea was not that women would fundamentally reform men, but that they would avert the worst male excesses and clean up the broken pieces. Thus the good works of one would cancel out the evil works of another.

Given this legacy of hypocritical and oppressive trade-offs, we should take special care that the cultivation of "feminine" ecological virtues not be abused as an excuse to continue other harmful habits. No matter how well intended the virtue ethic, its effects will surely be nullified. For example, when a chemical company takes certain "soft" (read: "feminine") measures, such as adopting an environmental ethics committee or initiating a tree-planting program, while leaving its "hard-core" (read: "masculine") production program unaltered, the ecological balance sheet will still not show a gain in clean air. Moreover, such calculated compensation strategies enhance moral hypocrisy and cynicism, which are dangerous attitudes in an ecological age.

(3) *The Escapism Test.* The cultivation of feminine virtues has also traditionally been encouraged to ensure the creation of pleasant escapes from the hardness of the world. Love, beauty, and understanding, cultivated by women, would enable tired men to recuperate at the end of an ugly workday.[38] While women were forcibly pushed into romantic roles, their very effectiveness in fulfilling these roles made it possible for the hardness of the world to continue—a hardness that not only played havoc with people, but also with the nonhuman world.

Such a romantic line of thinking forms another trap that must be avoided as we cultivate feminine ecological virtues. While ecoactivists may not be liable to fall into this trap, ordinary concerned citizens are. For example, in my native country, the Netherlands, a growing popular desire to care for nature has recently been combined with a growing need to seek respite from the stresses of modern life, producing a veritable national gardening craze.[39] Yet while shuffling in the backyard calms the uneasy consciences and re-energizes the bodies and souls of millions of citizens, the mushrooming "garden centers" make inordinate profits and the spirit of capitalism smiles once again. Except that Holland now has a lot fewer plain lawns and a lot more exotic plants that get replaced according to the latest fashion in garden color, not much has changed. Romantic escapism ultimately supports the status quo.[40] Again we see how easily the cultivation of feminine virtues such as caring and nurturing can be co-opted to serve reactionary ends. And this is still a benign example. As Chaia Heller reminds us, during the Nazi period, "Germans were encouraged to recover their close tie to nature, to the German wilderness, in order to purge themselves of the poison of foreign, decadent influences such as Leftists and Jews."[41] Love of nature lies dangerously close to love of nation. We should permanently guard against such escapist distortions of "feminine" virtue.

(4) *The Pedestal Test.* Many women have experienced the curious phenomenon of variably being praised and despised for the cultivation of feminine virtues, of alternately being placed on a pedestal and being dragged through the mud. This experience results from the schizophrenia of a patriarchal value system in which male alienation is expressed through both idolization and fear of females.[42] An article by ecofeminist Sharon Doubiago

nicely illustrates the situation. At one point Doubiago describes a typical pedestal, consisting of traditionally feminine virtues. The pedestal also reflects the desires for compensation and escape mentioned above:

> Women have had the role of, have been loved and valued for being, the Other—that is, the Voice of the heart, love, conscience, unconsciousness, compassion, sensitivity, sensuality, nature, nonlinear intuitive perception—the Other that men have demanded and greatly needed to turn back to, to return home to, from their ruthless world-making, their bloody battlefields.[43]

Elsewhere in the article, Doubiago shows how one of these "feminine" virtues, intuitive perception, is also regularly devalued in our society. She writes, "Woman traditionally listens to her inner voice. The habit is considered symptomatic; indeed, hormonal, anatomic."[44] The moral ambivalence expressed by these two passages is paradigmatic for the double standard operative in gender-constructed virtue language.

The persistence of patriarchal pedestals undercuts the effectiveness of any ethic in which "feminine" virtues play an important role. This is especially true for an ecological ethic, which is additionally hampered by the fact that nature tends to be put on the same pedestal as women. In order to be effective, an ethic needs to be wholeheartedly respected. This does not mean that we should surround "feminine" ecological virtues with artificial pomp and propaganda. (If mud is not dirty, then pedestals are unnecessary!) But it does mean that we should consistently expose any contempt that gets expressed for so-called "soft" ecovirtues,[45] and that we should continue to chip away at the patriarchal roots of this contempt.[46]

(5) *The Counterfeit Test.* Not all traditionally feminine virtues are genuine virtues; some are merely counterfeits, since they categorically block the flourishing of women (which surely cannot be reckoned as part of "the good"—whichever good!). Purity, for example, is an ideal of questionable value. As Catharina Halkes points out, feminine purity must be interpreted as the projection of a romantic, androcentric desire for escape.[47] In their attempts to cater to this desire, many women have ended up denying their own bodies, characters, and power. Similarly, obedience, as a "feminine" virtue, has meant little more than blind obedience to male commands (see Eph. 5:22–23!),[48] again an ideal of questionable value that has effectively oppressed many women in the past.

As ecologically minded people, we can learn from the widespread suffering caused by the cultivation of counterfeit "feminine" virtues. Although we may, with good reason, advocate caring for wetlands, compassion for animals, or sensitivity to the limits of ecological systems, we may want to think twice about advocating "back-to-nature purity," or "obedience to Nature." If historically the ideals of purity and obedience have been so consistently linked to the oppression of women, then it is quite possible that, even in ecological garb, these "virtues" are more likely to oppress people than to liberate nature.[49] A viable ecological virtue ethic should be clearly distinguishable from ecofascism, and listening to the voices of the oppressed is one way to do that.

CONCLUSION

At this point we may well raise a critical question: If the retrieval of traditionally feminine virtues for an ecological virtue ethic requires so much work, and if traditionally masculine virtues present even greater problems, given that they have been implicated in contributing to the current crisis, why should we bother with virtue language at all? Could we not do without virtue language altogether and just build an ethic around rights, duties, rules, or principles?

I do not think this is a realistic option. Any viable ethic must have a practical base, that is, roots in the ordinary language of people committed to causes other than devising ethical theories. If a theory ignores its base, it will itself be ignored. The practical base of a viable ecological ethic is the environmental movement. Here we find traditionally feminine virtues already well established, including in the language of ecofeminists. To dream of constructing an ecological ethic without terms such as love, care, compassion, humility, and sensitivity has at this point in time little practical value. We have no realistic alternative, then, but to proceed with these "feminine" virtues. However, we do have the option to proceed with caution. Historically speaking, the cultivation of "feminine" virtues has not improved the lot of women in Western society. The cultivation of these virtues in the context of an ecological ethic may, for parallel reasons, not improve the lot of nonhuman nature either.[50] Yet the predicaments of the past can also show opportunities for the future. If we are willing to learn from women's experiences of oppression, we will come closer to building a fully liberating ecological virtue ethic.

NOTES

1. Ecofeminists explore the symbolic and historical connections between the oppression of women and nature while seeking avenues toward the liberation and flourishing of everything within the web of life. Ecofeminist thought has multiple roots and these are a matter of some debate. Rosemary Ruether has described ecofeminism as a merger between feminism and certain schools of ecology, particularly deep ecology in the United States ("Ecofeminism: Symbolic and Social Connections of the Oppression of Women and the Domination of Nature, in *Ecofeminism and the Sacred*, ed. Carol Adams [Maryknoll: Orbis

Press, 1992]). Others see it simply as a logical outgrowth of feminism, particularly mediated through feminist pacifism (see Ynestra King, "The Ecology of Feminism and the Feminism of Ecology," in *Healing the Wounds: The Promise of Ecofeminism*, ed. Judith Plant [Philadelphia, PA: New Society, 1989], 18–28).

In what respect the method of this chapter can be characterized as ecofeminist is best indicated by Deborah Slicer's observation that "ecofeminists, as other feminists, trust the empirical facts of women's historical experience in order to formulate and to verify their theoretical work, and their theoretical work is not carried out simply to satisfy their intellectual curiosity, but to change the concrete empirical realities of life for those who suffer the oppression of 'patriarchal hell' " ("Is There an Ecofeminism-Deep Ecology 'Debate'?" *Environmental Ethics* 17 [1995], 163).

2. See, for example, Mary Daly, *Pure Lust: Elemental Feminist Philosophy* (San Francisco: Harper, 1984), 197–226; Anne Patrick, "Narrative and the Social Dynamics of Virtue," in *Changing Values and Virtues*, eds. Dietmar Mieth and Jacques Pohier, Concilium 191 Special Column (Edinburgh: T. & T. Clark, 1987), 69–80; Sharon D. Welch, *A Feminist Ethic of Risk* (Minneapolis: Fortress Press, 1990), 125–6 (Welch follows Michel Foucault here).
3. Mary Daly, "After the Death of God the Father: Women's Liberation and the Transformation of Christian Consciousness," in *Womanspirit Rising: A Feminist Reader in Religion*, eds. Carol P. Christ and Judith Plaskow (San Francisco: Harper & Row, 1979), 60.
4. Susan Moller Okin, *Women in Western Political Thought* (Princeton: Princeton University Press, 1979), 88.
5. Aristotle, *Nicomachean Ethics*, trans. W. D. Ross (Oxford: Oxford University Press, 1925), 1161^{a}20–25; see also Okin, *Women in Western Political Thought*, chap. 4.
6. Aristotle, *Nicomachean Ethics*, 1116^{a}10–15, 1150^{b}1–5.
7. Thomas Aquinas, *Summa Theologiae* II–II.151.1, trans. the Fathers of the English Dominican Province (New York: Benziger Brothers, 1948). Interestingly, this assumption occurs in a passage criticizing the hypocrisy of husbands who expect their wives to be chaste while they themselves yield "to the first onslaughts of lust." Augustine and Aquinas thus criticize one double standard on the basis of another double standard. Their criticism contributes to the internal consistency of gender-constructed virtue language; it must not be mistaken for a criticism of this language itself.
8. Ibid., II–II.128 *ad* 6.
9. See Jean Bethge Elshtain, *Public Man, Private Woman: Women in Social and Political Thought* (Princeton: Princeton University Press, 1981), 91–9.
10. The term *virtú* for Machiavelli simply means "manliness" (he thereby stays close to the etymological root of the term, which is derived from the Latin *vir*, man). For further analysis, see Hanna Fenichel Pitkin, *Fortune Is a Woman: Gender and Politics in the Thought of Niccolò Machiavelli* (Berkeley: University of California Press, 1984).
11. Riane Eisler, "The Gaia Tradition and the Partnership Future: An Ecofeminist Manifesto," in *Reweaving the World: The Emergence of Ecofeminism*, eds. Irene Diamond and Gloria Orenstein (San Francisco: Sierra Club Books, 1990), 30.
12. Catharina Halkes, *En Alles Zal Worden Herschapen: Gedachten over de Heelwording van de Schepping in het Spanningsveld tussen Natuur en Cultuur* (Baarn, The Netherlands: Ten Have, 1989), 60 (my own translation). See also the English language edition: Catharina J. M. Halkes, *New Creation: Christian*

Feminism and the Renewal of the Earth (Louisville, KY: Westminster/John Knox Press, 1991), 48.

13. See, for example, Aristotle, *Nicomachean Ethics*, 1140^{b}10–20; see also Thomas Aquinas, *Summa Theologiae* II-II.47.16, II-II.53.6.
14. Halkes, *En Alles Zal Worden Herschapen*, 60.
15. See Appendix A. Ecowriters also advocate traditionally "masculine" virtues, such as competence, courage, critical reflection, determination, discipline, efficiency, humor, justice, leadership, managing/stewardship, perseverance, prudence, realism, responsibility, self-confidence, self-reliance, and vigilance. For a discussion of ecological courage, see chapter 8.
16. The first, second, and fourth options parallel Ynestra King's typology for conceiving the woman-nature connection. See King, "The Ecology of Feminism and the Feminism of Ecology," 22–3.
17. Josephine Donovan, "Animal Rights and Feminist Theory," in *Ecofeminism: Women, Animals, Nature*, ed. Gaard (Philadelphia, PA: Temple University Press, 1993), 167.
18. Tom Regan, *The Case for Animal Rights* (Berkeley and Los Angeles: University of California Press, 1983), xii (cited in Donovan, "Animal Rights and Feminist Theory," 167–8).
19. As Linda Vance argues, "Identification and empathy may be dismissed by rationalists as sentimental—as feminine—but passionate convictions, beliefs from the heart, can always get us through the hard times when reason and argument fail" ("Ecofeminism and the Politics of Reality," in *Ecofeminism: Women, Animals, Nature*, ed. Gaard, 136).
20. On the symbolic connection between woman/nature as contrasted to man/culture see, for example, Sherry B. Ortner, "Is Female to Male as Nature Is to Culture?" in *Woman, Culture, and Society*, eds. Michelle Rosaldo and Louise Lamphere (Stanford: Stanford University Press, 1974), 67–87.
21. As Lori Gruen observes, radical feminists "embrace the connection [between woman and animals/nature] and attempt to strengthen it by denying the value of its opposite. In other words, radical feminists see women as closer to nature and men as closer to culture and thereby reject the cultural in favor of the natural. They elevate what they consider to be women's virtues—caring, nurturing, interdependence—and reject the individualist, rationalist, and destructive values typically associated with men" ("Dismantling Oppression: An Analysis of the Connection Between Women and Animals," in *Ecofeminism: Women, Animals, Nature*, ed. Gaard, 77). For a concrete example of women redefining and affirming the woman-nature connection, see Vandana Shiva, *Staying Alive: Women, Ecology and Development* (London: Zed Books, 1989), 47.
22. Ynestra King, Janis Birkeland, and others make the additional argument that "since all life is interconnected, one group of persons cannot be closer to nature" (Janis Birkeland, "Ecofeminism: Linking Theory and Practice," in *Ecofeminism: Women, Animals, Nature*, ed. Gaard, 22). Along the same lines, Joan Griscom suggests that "'Only the nature/history split allows us even to formulate the question of whether women are closer to nature than men. The very idea of one group of persons being "closer to nature" than another is a "construct of culture"'" (Ibid., 22). I do not find these arguments particularly helpful, since they obscure the question of the quality of one's connection with the nonhuman world. While I agree that we are all connected to nonhuman nature, some of us are better attuned than others.

23. Gruen, "Dismantling Oppression," 80. In general, Gruen highlights several of the same problems addressed in this paragraph (see esp. 77–8).
24. This is not a necessary implication, since one may hold that virtues are *cultivated* qualities that require no natural affinity.
25. Some ecofeminists have pointed out that the environmental movement is currently segregated, with women being the second-class citizens. See, for example, Linda Vance, "Ecofeminism and the Politics of Reality," 124.
26. This argument is a mirror image of the common ecofeminist claim that male exclusivism toward women has spilled over into their attitudes toward the non-human world.

    Personally, I see support for the argument insofar as I believe that the virtue of inclusivism cannot be fully cultivated if one practices exclusivism with respect to any particular party. Exclusivism implies that judgmentalism is part of one's character. Thus one's inclusivism toward certain parties is by definition *conditional*. The possibility of rejection always lurks closely under the surface. Such selective inclusivism is only a semblance of the true virtue, which is based on compassionate understanding. (Incidentally, compassionate understanding can be critical of what is harmful, but that would never cause it to *reject* another party.)
27. The "Head-on Option" may also create hypocrisy if it is not accompanied by sufficient self-examination. Josephine Donovan warns that "one cannot simply turn uncritically to women as a group or to a female value system as a source for a human relationship ethic with animals," because women "have been complicit in that abuse, largely in their use of luxury items that entail animal pain and destruction (such as furs) and in their consumption of meat" ("Animal Rights and Feminist Theory," 168). Such hypocrisy will surely not be missed by the media either.
28. Birkeland, "Ecofeminism: Linking Theory and Practice," 26.
29. On the inappropriateness of "color blindness" in business, see Lennie Copeland, "Learning to Manage a Multicultural Workforce," *Training* 25 (May 1988), 48–9, 51, 55–6.
30. Phyllis Trible proposes a similar approach in dealing with patriarchal texts in the Bible. See *Texts of Terror* (Philadelphia: Fortress Press, 1984).
31. Birkeland, "Ecofeminism: Linking Theory and Practice," 17, 20.
32. This implies that we take another look at so-called "masculine" ecological virtues as well. The field of men's studies could help in sifting the more helpful from the more harmful elements in these virtues. I also think we should more critically examine the commonplace rejection of "masculine" vices such as aggression and rationalism. It would be helpful if we saw these along the same lines as Aristotle sees ambition: harmful if too extreme, but helpful in the right amount for a particular context. Here too I think it is important to keep the question of a genetic basis open for inquiry rather than dogmatically decided.
33. Vance, "Ecofeminism and the Politics of Reality," 124.
34. For a related argument, see Rosemary Ruether, *New Woman/New Earth: Sexist Ideologies and Human Liberation* (New York: Seabury Press, 1975), 200–4.
35. When businesses adopt recycling and ride-share programs without substantially altering their production and distribution patterns, this does not change the situation of inequity. Such businesses simply bring the private/public split within their own walls, creating a hypocrisy that is very obvious to employees and the general public. Still, even limited efforts such as recycling should be com-

mended; they may have subversive effects. My business students tell me stories of employers who find that the daily recycling of office paper has caused a mushrooming environmental concern among their employees, accompanied by demands for further changes.

36. Consider, for example, that municipal solid waste, upon which homemakers and cottage industry owners can have an impact, constitutes only 1 percent of U.S. annual solid waste production (see Walter H. Corson, ed., *The Global Ecology Handbook* [Boston: Beacon Press, 1990], 267).
37. Halkes, *En Alles Zal Worden Herschapen*, 60.
38. Cf. Ruether, *New Woman/New Earth*, 196–200.
39. Annegreet van Bergen and Marijke Hilhorst, "De Lusthof Nederland," *Elsevier* 25 (22 June 1996), 64–70.
40. Along the same lines Janis Birkeland observes: "Mainstream emphasis on the individual 'at one with nature' distracts attention from structural and systemic issues. Institutions embody values, so they must be changed as well" ("Ecofeminism: Linking Theory and Practice," 45).
41. Chaia Heller, "For the Love of Nature: Ecology and the Cult of the Romantic," in *Ecofeminism: Women, Animals, Nature*, ed. Greta Gaard, 237.
42. For a nice visual representation of the pedestal treatment, see *Tussen Heks en Heilige* [*Between Witch and Saint*]: *Het Vrouwbeeld op de Drempel van de Moderne Tijd, 15de/16de Eeuw* (Nijmegen: Uitgeverij SUN, 1985).
43. Sharon Doubiago, "Mama Coyote," in *Healing the Wounds: The Promise of Ecofeminism*, ed. Plant, 41 (inspired by Dorothy Dinnerstein, *The Mermaid and the Minotaur*).
44. Ibid., 40.
45. Just as, when push comes to shove, the EPA is not taken as seriously as the Pentagon in policy formation, so many people still do not take caring for the environment as seriously as, say, being a "tough" manager at work—much lip service to the contrary.
46. Janis Birkeland suggests, for example, that "we must expose the assumptions that support Patriarchy and disconnect our concept of masculinity from that of 'power over' others and the rejection and denigration of the 'feminine'" ("Ecofeminism: Linking Theory and Practice," 19).
47. Halkes writes, "Women are always evaluated from an androcentric point of view: be it as inferior, as a mere body with emotions and passions, or as an instrument for procreation and for the care of daily household life; or be it as exalted, as Muse, as guardian angel, as the Virgin Mary, pure and chaste, as 'The Lady,' as the poetry in the daily prose, and thus also as a means of escape" (*New Creation*, 48). Both the trade-off mentality and the pedestal idea shine through in this passage.
48. Susan Griffin cites this example in *Woman and Nature: The Roaring Inside Her* (New York: Harper & Row, 1978), 10.
49. Note, in this context, the parallel between instrumentalist, anthropocentric attempts to keep natural wilderness "pure" as a place of retreat for humans, and attempts to con women into providing retreats for men (see Halkes, *En Alles Zal Worden Herschapen*, 59–60).
50. An even stronger claim can be made if one takes as a starting point the thesis, supported by many ecofeminists, that human domination of nature is rooted in male domination of women (see, for example, King, "The Ecology of Feminism," 24). If this is true, then gender-constructed virtue language, which has

sustained the oppression of women, must itself be considered a contributing factor in the network of causes underlying the ecological crisis. The implication is that retrieval of gender-constructed virtue language for an ecological age will *ipso facto* be counterproductive, unless all gender constructions can be undone.

Personally, I prefer to take an inductive approach that looks at individual gender-constructed virtues and makes projections on the basis of concrete problems associated with these virtues in the past. I present my conclusion as the sum of such projections. This approach has the benefit of circumventing problems associated with demonstrating the thesis that the domination of nature is rooted in the domination of women.

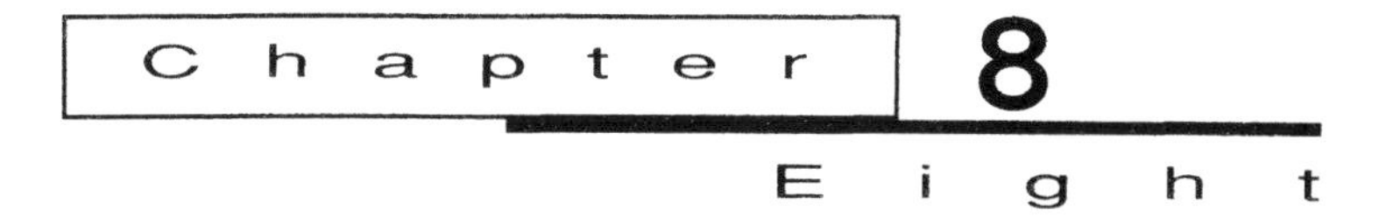

SUSTAINABLE VIRTUE: A PSYCHOLOGICAL ANALYSIS OF COURAGE

IN THE PREVIOUS CHAPTER I showed how women's experiences of oppression can become sources for critical reflection on ecological virtue discourse. Another promising source is the language of psychology. With notions such as internalization, repression, and projection, psychological language offers insights into the limits and opportunities of human moral behavior that were unavailable when Western virtue traditions were shaped.[1] Here we find a source that warrants independent attention.[2]

In order to illustrate how psychological insight can help determine when human dispositions and actions warrant the label "genuinely virtuous" in an age faced with serious ecological problems, I will focus on the virtue of courage. This traditionally masculine virtue occupies an important place in ecological virtue discourse. Courage is needed to shake familiar, but unsustainable habits and to challenge ecologically harmful practices, institutions, and structures of power. Moreover, courage is needed to venture into the unknown, to make new beginnings.[3] At a more general level, courage is a necessary prerequisite for the cultivation and practice of other ecological virtues. Without courage, one would not have the ability to persist with good habits such as frugality and temperance in a world that is likely to welcome such habits with mockery and threats (stories abound that—sometimes tragically—illustrate this point). It is hardly surprising, then, that ecologically minded people recognize courage as an important virtue.

We may well ask, however, how courage is best perceived and cultivated in an ecological age. In this chapter I argue that current popular conceptions of courage—what Lee Yearley has called "Rambo-like pictures of courage"[4]—are more harmful than helpful. Unfortunately, such pictures belong to popular environmentalism as well. Chaia Heller offers a telling description:

> As we observed on Earth Day 1992, politicians, corporate ringleaders, and deep ecologists are leaping into the romantic, ecological drama, becoming "ecoknights" ready to protect and save helpless "Lady Nature" from the big, bad dragon of human irresponsibility.[5]

I propose that Rambo-like pictures of courage generate psychological chain reactions that lead to complacency, grandiosity, and a compulsion to control, all of which are ecologically harmful attitudes. Thus, any thorough and consistent effort to overcome ecologically harmful attitudes will have to include a critical assessment of the popular perception and practice of courage.

In this chapter I offer such an assessment, again in dialogue with the Aristotelian-Thomistic virtue tradition. But my main criterion is inspired by psychological discourse, namely, that in an ecological age the cultivation of courage—and of any virtue for that matter—needs to be *sustainable*, in the sense that it should not simultaneously generate ecologically harmful side effects through the dynamics of repression and projection.[6] In other words, I propose that we approach the cultivation of character along the same lines as we are (re)learning to approach the cultivation of land.

Popular Courage

Virtue theorists generally agree that courage is essential for any human society to function well.[7] Western virtue tradition has acknowledged this insight by calling courage a "cardinal" virtue, that is, a virtue on which human life "hinges."[8] Especially the idea that courage is essential for the preservation of human reasoning capacities has been emphasized, which is why courage has also received the title of "preservative virtue" (along with temperance, even-temperedness, perseverance, and patience).[9] Indeed, it is difficult to imagine how humans could live as purposeful agents without some way to prevent fear from deflecting their plans.

To say this, however, is not to say very much. There are many possible ways for humans to experience fear, to relate to their plans, and to keep fear from interfering with their plans. As a result, different cultures have developed significantly different interpretations of courage, which cannot simply be "stripped" to reveal a substantive cross-cultural "core" of what courage means. In other words, the details of these culturally embedded interpretations are essential, and any meaningful discussion of a particular concept of courage will have to rely on "thick description."[10]

When we turn to the popular "Rambo-like pictures of courage," thick description reveals at least two core assumptions, namely, that "fear is not okay" and, derivatively, that "fear should be controlled." I will show by means of psychological inferences how both of these assumptions are problematic in an ecological age.

At a popular level, a major test of character is whether or not one experiences fear. The ideal hero is never afraid. The less ideal hero may look more human for showing fear in a weak moment, but he also risks looking like a sissy. No matter how good one is at controlling fear, it would still be better not to experience it at all.

Unfortunately, the belief that "fear is not okay" provides a perfect breeding ground for psychological repression. Fear is an unavoidable human emotion. When an unavoidable emotion becomes culturally unacceptable, people who experience it anyway will likely deny and suppress it. In the case of fear, the potential effects of such a reaction are particularly worrisome in an ecological age. Repression leads to contempt, both for the part of oneself that is not accepted, and for others who manifest what one does not accept in oneself.[11] Such contempt can find expression in the form of two harmful attitudes, namely, complacency and grandiosity.

The complacent person denies that there might be legitimate grounds for fear by denying that there would be grounds for action. In the interest of keeping internal peace, she will construe a comforting picture of reality that will make both fear and action look unnecessary. At the same time, she is liable to judge those who call for action as alarmist extremists. Overall, we can say that the complacent person displays armchair courage, claiming that she would be courageous if it were necessary, but construing reality in such a way that she will rarely consider courageous action as warranted.

The dangers of such judgmental complacency in an ecological age should be obvious. On the one hand, we are confronted daily with warning signals of environmental destruction—signals that, if taken seriously, should inspire genuine fear. On the other hand, it is easy to downplay the significance of these signals, for all too often they seem remote in distance and time and therefore unreal (such as ozone depletion, soil erosion, aquifer deterioration, etc.). Those who cannot admit to fear because it would lead to self-contempt are likely to capitalize on the remoteness of the evidence and delay action at a time when turnaround options are rapidly diminishing. In sum, complacency is ecologically harmful in two ways: It can obstruct the action of others who are judged extremist, and it prevents the self from acting when action may be necessary.

By contrast, the grandiose person does see grounds for action, but he too lives by the assumption that "fear is not okay" and will consequently deny that fear is among his motives for action. Thus emerges the "fearless" hero (the "Rambo" type impersonated). In such a person, the self-contempt resulting from the repression of fear finds expression through compensation: By adopting a heroic posture he can have a socially acceptable self-image that provides credible motives for action other than fear (e.g., he can claim a desire to help, or to do noble things). In other words, grandiosity appears as a side effect of the self-contempt generated by the repression of fear.[12]

In an ecological age such grandiosity is particularly worrisome, since it feeds into an attitude that is widely regarded as very harmful to the environment: arrogance. The grandiose person sees himself as better than others, including those who benefit from his courageous action.[13] Thus the grandiose hero who turns his attention to saving the environment may subconsciously perceive the environment as a helpless victim. This makes his actions self-defeating, for the arrogance with which he acts belongs to the same hierarchical pattern of relating that, according to many ecowriters, has contributed to the ecological crisis in the first place.[14]

Not everyone who has been shaped by the idea that "fear is not okay" will resort to repression, however. An alternative strategy for dealing with socially unacceptable fear is offered by the second assumption revealed by a thick description of popular courage, namely, that those who cannot avoid feeling fear should keep it under control. Unfortunately, though, this strategy can also generate side effects that are problematic from an ecological perspective.[15]

There are various ways to control fear, the crudest one being a constant, watchful constraint of the emotion, as if one were harnessing a wild beast. This approach, which I will refer to as "heavy-handed control," leads at best to an unstable situation: One may feel in charge of one's fear, but one can never relax lest the fear "escape" and take over. This instability is unsettling and may create a compulsive need for reassurance that one is going to stay in charge. People who suffer from such a compulsion tend to use their surroundings to achieve the sense of reassurance they need. They seek to dominate their surroundings, making visible the kind of successful control that they so depend on for their inner peace. (For example, concentration camp survivors may develop a compulsion to maintain an impeccably organized household.) Thus, greater external control provides greater inner security for those who are always uncomfortably keeping their fear in check.

One arena in which people have sought to exercise their compulsion to dominate is nonhuman nature. In Western history, the development of the tendency to gain internal security through external control of nature can be traced through changes in the medieval iconography of the virtue of temperance (which, as a preservative virtue, resembles courage in many respects). As Carolyn Merchant points out in her book, *The Death of Nature*:

> The figure of Temperantia evolves from a picture of a woman pouring water into a jug of wine, to one holding a pair of compasses (1359) and an hourglass (1350s). By the early 1400s, she is shown adjusting a mechanical timepiece and soon after is surrounded by geared clocks and other measuring and surveying instruments of great intricacy (ca. 1450).[16]

Temperantia, in other words, exchanged the attributes of personal control for the attributes of control over nature. This exchange, I submit, is not accidental, but the result of the human tendency to control the outer world as a response to the instability created by heavy-handed inner control (during the period described by Merchant, this tendency may have been encouraged by fear of the plague).

When the need for external control is turned toward the natural environment, it can create serious problems. Whereas for centuries it was thought appropriate that human beings would aim at attaining "mastery" over nature, many modern ecowriters point out that attempts at mastery and control over nature more often than not have destructive results.[17] We have good reason, then, to be suspicious of the controlling tendencies of those who subject themselves to the heavy-handed internal control of fear.

Thus, a thick description of the popular conception of courage yields at least three potential connections between the cultivation of courage and the emergence of ecologically harmful attitudes. Complacency, grandiosity, and a compulsion to control are all integrally linked to basic assumptions about fear that underlie the popular conception of courage. As long as this conception continues to inspire, it will continue to undercut attempts to redefine human relations with nonhuman nature. In other words, the popular conception of courage is unsustainable in an ecological age.

At the same time, however, courage in some form or another is essential for the functioning of human society. The question before us, then, is how we can conceptualize courage such that the psychological generation of ecologically unsustainable attitudes can be avoided.

Courage in an Ecological Age

In order to know how courage can be sustainably cultivated, we first need to know how we can avoid the factors that make its current popular cultivation unsustainable. What can we do to prevent the psychological chain reactions that give rise to complacency, grandiosity, and the compulsion to control? I propose three requirements, each of which involves the cultivation of a virtue: earthiness, imaginative channeling, and vulnerability. In other words, these three virtues can be seen as constituent parts of genuine ecological courage.[18] Unlike the Rambo-type cultivation of courage, they are consistent with ecological aims and sensibilities. They are also consistent with some aspects of the Aristotelian-Thomistic virtue tradition, but they divert from certain more patriarchal aspects of this tradition.

In the previous section I argued that the psychological chain reactions that end in ecologically harmful complacency and grandiosity find their roots in the assumption that "fear is not okay," an assumption that belongs to the popular Rambo conception of courage. Taking the argument one step further along the line of psychological inferences, I now propose that this assumption is both the result of and itself results in a strong sense of shame. On the one hand, the idea that "fear is not okay" expresses a deeper sense that fear is shameful (for example, because it shows "weakness," or "feminity," or "lack of nobility"; here the patriarchal character of traditional courage shows very clearly). On the other hand, once a person accepts the idea that "fear is not okay," any experience of fear will arouse a sense of shame, which that person may seek to eliminate through denial and suppression.

It seems, then, that the key to avoiding the ecologically harmful attitudes of complacency and grandiosity is not to be subject to feelings of shame in connection with the experience of fear. I propose that we can achieve this if we can wholeheartedly accept fear as a natural and often useful reaction to perceived danger. Such acceptance requires that we be comfortable in thinking of ourselves as animals. After all, we share the experience of fear with other animals, and as long as that thought makes us uncomfortable, we will be ashamed of our fear. Another way of putting this is that the acceptance of fear requires us to have a certain earthy attitude toward ourselves, by which I mean that we see ourselves realistically (in a down-to-earth manner) and appreciate our physical and animal characteristics.

This proposal signifies a move away from the deeply rooted Western tradition that links courage with the ideal of nobility. Aristotle, for example, argues that "death and wounds will be painful to the brave man and against his will, but he will face them because it is noble to do so or because it is base not to do so."[19] Baseness, for Aristotle, connotes being and acting like an animal, whereas nobility connotes having cultivated *distinctively* human characteristics, that is, characteristics we do *not* share with other animals.[20] Clearly then, the traditional argument that nobility is the only proper motive for courage runs exactly counter to the idea that courage requires our being comfortable with our animal characteristics.

By contrast, as I have pointed out in chapter 2, ecologically minded authors are already recommending precisely such an earthy attitude. Ecoliterature is filled with phrases such as "those who fully accept themselves as part of the natural world,"[21] and "the natural world is the maternal source whence we emerge into being as earthlings."[22] In some of the literature the attitude of earthiness even takes on a spiritual dimension when the constitutive experience of kinship with other creatures and with the Earth is described in terms reminiscent of religious experience.[23] This earthy attitude includes the acceptance of fear, even the terrible fear we may experience upon realizing the extent of the ecological crisis. As Roger Gottlieb observes, "There is nothing shameful or 'weak' in the pain we feel about the environment. Grief and fear are rational responses to our losses and perils."[24] In fact, Gottlieb and others point out, the first step toward healing and effective action is to admit our fear.[25]

To recap then, through psychological reasoning we can trace complacency and grandiosity to the dynamics of shame. These dynamics can be prevented by an earthy attitude through which we accept fear as a normal animal reaction to perceived danger. The promotion of such earthiness would be consistent with existing ecological virtue discourse, but it would go against Aristotelian tradition. In this situation of conflicting moral sources I propose that, if we are genuinely interested in building an ecologically sustainable life of virtue, we should be prepared to challenge tradition on the basis of psychological and ecological insights and accept earthiness as one of the constituents of ecological courage.[26]

Two more constituents—imaginative channeling and vulnerability—can be identified by examining how the Rambo conception of courage generates the compulsion to control the external world. I have argued that this compulsion results from the assumption that fear requires heavy-handed inner control. The key, then, to avoiding the compulsion to master the environment would be to drop this assumption and to look for alternative ways of dealing with fear. I will identify two complementary alternatives.

Just as a "naughty" child does not necessitate spanking, a surge of fear does not necessitate militaristic control. Yet it requires imagination (and perhaps the help of a psychologist) to think of other options. The parents of the "naughty" child may, to their surprise, discover that offering the child a positive distraction, or giving him or her some quality attention, is a viable way of dealing with the problem. Similarly, we can reduce the unsettling effects of fear by sharing it with others, or by thinking of worst-case scenarios and how we would cope in those situations, or by doing something practical that reduces the danger.[27] These and other similarly creative options have the effect of giving us a greater sense of inner harmony, achieved through channeling fear, rather than through harnessing it. In other words, they allow the emotion to run its course within acceptable limits, rather than capping it, which is always potentially explosive. And because imaginative channeling enables greater internal harmony, it is unlikely to have the negative side effect of a compulsion to control the external environment.[28]

To some extent, the idea of imaginative channeling stands in tension with the Aristotelian-Thomistic virtue tradition, but it could also consistently be interpreted as an extension of this tradition. The idea that one needs to "master fear"[29] has a long history of support, yet the tradition does not seem to provide a binding reason explaining why fear calls for heavy-handed control. The idea seems to be based primarily on popular assumptions. At the same time, imaginative channeling is consistent with the tradition, though it implies an extension. It is consistent because the tradition does make allowance for appropriate levels of fear and does not consider these shameful (unlike popular culture). Both Aristotle and Aquinas argue that the person who truly values his life will feel an appropriate amount of fear.[30] But if fear is acceptable, then presumably imaginative ways of dealing with it that do not aim at forcing it away would be acceptable too.

Imaginative channeling also fits within an ecological perspective. Ecowriters frequently call for and propose imaginative solutions to problems that traditionally have been thought to require a controlling response. Organic alternatives to pest and weed control provide just one example. Such practical alternatives are not only about imagination, but also about ways of dealing with fear. Reliance on chemical pesticides and herbicides is a direct response to fear, namely, fear of losing a harvest and perhaps one's livelihood. Ecowriters show us that this fear need not be met with overkill (literally), but that it can be channeled and reduced through increasing our empathic understanding and ability to work within ecological systems. Personally, I still remember the day when my aunt, who for many years had worked at the New Alchemy Institute on Cape Cod, told me that her aim was not to grow hole-free cabbages, but simply to grow enough cabbage for us to eat. It felt like a revelation to realize that we could *share* crops with "pests," rather than always feel afraid of them.

In sum then, the ecologically harmful compulsion to control, which appears as a side effect of the popular Rambo conception of courage, can be avoided through imaginative channeling of fear. The promotion of such channeling is consistent with existing ecological virtue discourse, as well as with Aristotelian-Thomistic virtue tradition. On all accounts, then, it makes sense to accept imaginative channeling as one of the constituent elements of sustainable ecological courage.

Imaginative channeling itself, however, is based on yet another psychological precondition, namely, vulnerability. In order to let fear run its course within acceptable limits, we need to be able to go *through* the experience, rather than run from it. After all, fleeing in times of fear is the mark of a coward, and courage requires precisely that we can stand fast to defend what we cherish. There is no way around it: True courage must somehow involve the ability to embrace fear (which, by the way, does not imply enjoyment). This again requires a basic personal attitude, namely, vulnerability. People who both *accept* their existential vulnerability and can *make* themselves vulnerable (i.e., open) will have the ability to experience fear without panic.[31] This will enable them to respond to dangerous situations with maturity and without harmful side effects.

The requirement of vulnerability is consistent with tradition also. Both Aristotle and Aquinas assume that the truly courageous person has the ability to face fear head-on. By contrast, those who do not fully face fear—for example, those who downplay real danger or those who are overconfident—are not truly courageous.[32] Josef Pieper offers a modern version of the same line of thought by arguing that fortitude always involves the acceptance of insecurity.[33]

Vulnerability is also an important concept for ecowriters. Both Rosemary Ruether and Dorothee Soelle, for example, advocate vulnerability as the only way to true peace.[34] Vulnerability in the context of an ecological worldview implies the ability to face our creaturely limits, especially death, and to accept our dependence on the web of life.

In sum, psychological insight, tradition, and ecological discourse converge on the importance of vulnerability in the face of fear. On the basis of this convergence we can reasonably conclude that such vulnerability forms a third constituent of sustainable courage. In other words, the cultivation of courage in an ecological age should involve the cultivation of the ability to be vulnerable.

Earthiness, imaginative channeling, and vulnerability help us avoid some of the environmentally harmful attitudes generated by popular ways of dealing with fear. They make courage a sustainable option.[35] At the same time, these dispositions allow us to deal with fear in such a way that we can continue on our life course without being paralyzed by fear. Specifically, they allow us to perceive and do what needs to be done for the flourishing of the biosphere, even in the face of personal danger. Earthiness, imaginative channeling, and vulnerability make for true courage.

The Perfection of Courage

One difficulty remains, however: Fear feels unpleasant and courage is therefore never a psychologically stable virtue. The only solution to this difficulty would be a radical emotional transformation that eliminates the experience of fear without requiring repression or a misjudgment of the degree of danger.[36] Such a transformation, however, is not fully a matter of will. Hence it can only be reckoned as belonging to the perfection of courage; it is not a constituent of sustainable courage. Yet it would provide the best guarantee of sustainability. On this point, Thomas Aquinas's reflections on courage actually provide a remarkable degree of psychological insight.

Even with earthiness, imaginative channeling, and vulnerability, acts of courage remain uncomfortable experiences. No matter how we adjust our basic attitudes, in situations of danger we still face potential or actual pain or even loss of life. As long as that reality makes us afraid, we simply cannot feel too good. For this reason, I tend to think of courage as an "uneasy virtue"—it is never an attitude in which we can feel at home.

As a result, the practice of courage will constantly lead us to the edge of developing a need to compensate for our fear. Too much un-at-homeness may, for example, turn us into romantics who desperately long for experiences of total at-homeness (witness the legendary appeal of a woman's embrace to a soldier during wartime). To the extent that we turn this romantic desire toward nature, we may again be undercutting our very efforts at rebuilding right relationships with the nonhuman world. The back-to-nature urge, the desire for total harmony with nature, may well get in the way of establishing mutually helpful patterns of interaction.[37] Thus the uneasiness of courage presents us with an unrelenting challenge that is unavoidably linked to the nature of courage.

There is a way, however, to face this challenge. But it involves a transformation of the nature of courage into an approach to danger that can only by analogy be called courage. In his book, *Mencius and Aquinas*, Lee Yearley describes Thomas Aquinas's understanding of such a transformation of courage from an uneasy virtue into a virtue marked by spontaneous ease. Aquinas attributes the change to the spiritual Gift of Courage:

> Aquinas speaks here about a state of being in which the Holy Spirit's actions generate a confidence that excludes fear and leaves a person completely tranquil. The direct action of deity produces a state where a preservative virtue becomes an inclinational virtue. Adverse movements no longer need to be resisted; only animating inclinations are present.[38]

Such a transformation would effectively remove ecological courage from the edge of danger, so we would do well to hope for it.[39]

As it stands, Aquinas's description of this process would not appeal to large numbers of ecologically minded people who consider themselves recovering Catholics, Methodists, and so forth. Given that virtues are always practiced in concrete communal contexts, and given that the environmental movement is the primary community for the cultivation of ecological courage, this presents a genuine problem. We can, however, conceive of the transformation of courage in a manner analogous to Aquinas's description, but with more general experiential appeal. Specifically, I propose that we think of the transformation of ecological courage as mediated through the experience of love (which some ecowriters would be prepared to link with the creative power of the divine).[40]

When the experience of love is powerful enough, it has the effect of gathering and focusing our energy, which otherwise would flow in many directions. As a result of this dynamic, we create such richness in one area of our lives that we feel virtually untouched by problems in other areas of our lives. Moreover, this richness may spill over into a desire to accept and embrace what would normally make us cringe and run. The experience of love, then, effectively changes our experience of fear.[41] This enables us to pursue justice calmly and happily, even in situations that would ordinarily inspire great fear. We could call such action courageous, but only by analogy, suggesting that *ordinarily* such action would require the ability to deal with fear.

This line of reasoning in some respects resembles Augustine's reinterpretation of the classical cardinal virtue of courage as "love enduring all things willingly for the sake of that which is loved."[42] Augustine refers to the psychological transformation occasioned by the experience of love, arguing that "there is nothing, however hard or unyielding, which cannot be conquered by the fire of love."[43] Yet he draws a sharp distinction between those who become courageous from loving earthly things, such as gold, praise, and women, and those who become courageous from loving God. The latter, he believes, will have more courage, for "when borne by love toward God, the soul will soar free and glorious above all torment on the beautiful and unblemished wings with which chaste love rises to the embrace of God."[44] Many ecologically minded people (though not all) would prefer not to draw such a sharp distinction between love of God and love of earthly things.

Courage in its most perfect and sustainable form, then, is marked by love. As such it no longer performs the traditional preservative function, since the threat against which the moral self needs to be protected, that is, the disturbing effect of fear, has been removed. Instead, transformed courage functions as an inclinational or substantive virtue, specifically, as a "form of benevolence" directly inspired by concern for the other.[45]

This, of course, is an ideal, since few people permanently find themselves in a state of consciousness that is marked by love.[46] Yet some reflection of the ideal is normally present in people we find truly courageous. As Robert C. Roberts observes, "it is typical of courageous people that they are passionate." In fact, Roberts considers "enthusiasm" a necessary requirement for virtues of willpower and argues that to some extent this is a skill that can be cultivated.[47] The ideal, then, is not entirely foreign to our ordinary experiences, nor should we simply passively hope for it to happen to us. Just as Aquinas's Gift of Courage requires a degree of spiritual readiness and openness on the part of the recipient, so we can prepare ourselves for the transformation of courage through love by actively creating and cultivating the conditions that make love possible.

CONCLUSION

An ecologically sustainable concept of courage includes the constituents of earthiness, imaginative channeling of fear, and vulnerability. In other words, the cultivation of courage in an ecological age requires the prior and deliberate cultivation of these related attitudes. The best guarantee of sustainability, however, comes from the transformation of courage from a preservative virtue into a virtue of benevolence. With these conclusions I intend not only to contribute to ecological virtue theory by proposing concrete features of an ecological concept of courage, but I also intend to show the value of paying attention to psychological discourse in the interest of protecting and fostering good human-nonhuman relations. Just as sustainable agriculture requires knowledge of the long-term and systemic effects of cultivation practices, sustainable virtue requires knowledge of the long-term psychological effects of the cultivation of character. Thus psychological discourse, as a systematic way of reflecting on human experience, deserves a place next to ecological virtue discourse, religious and philosophical tradition, and women's voices as a source worth consulting in the effort of shaping a viable ecological virtue ethic.

NOTES

1. Yet I agree with David W. Kidner that, in order for psychological discourse to contribute more fully as a source for environmental ethics, it needs to be itself transformed to include a greater historical and critical consciousness. See David W. Kidner, "Why Psychology Is Mute about the Environmental Crisis," *Environmental Ethics* 16 (1994), 359–76. For an interesting attempt to integrate psychology, ecological awareness, and virtue theory, see John H. Riker, *Human Excellence and an Ecological Conception of the Psyche* (Albany, NY: State University of New York Press, 1991).
2. The same could be said of other modern forms of reflection on human experience, such as sociology and social anthropology.
3. See, for example, Douglas C. Bowman, *Beyond the Modern Mind: The Spiritual and Ethical Challenge of the Environmental Crisis* (New York: Pilgrim Press, 1990), xiii. Bowman also sees learning as courageously being open to the new (36).
4. Lee H. Yearley, "Recent Work on Virtue," *Religious Studies Review* 16 (January 1990), 2.
5. Chaia Heller, "For the Love of Nature," in *Ecofeminism: Women, Animals, Nature*, ed. Greta Gaard (Philadelphia, PA: Temple University Press, 1993), 219.
6. Some of the "feminine" virtues that are now reclaimed for an ecological age also have a history of being cultivated in a psychologically unsustainable manner. A good example is the virtue of compassion. As Judith Plant points out:

   > Women have been socialized in such a way that allows them to experience compassion. This experience is, however often skewed by the subordinate, deferential position given to women that somehow loads the responsibility of caring for others with guilt and anxiety. For this compassion that women have been allowed does not carry with it any power to make decisions. Because of this skewed situation, caring often becomes entangled with personal frustrations over feelings of powerlessness, leading to an inability to take responsibility ("Toward a New World: An Introduction," in *Healing the Wounds: The Promise of Ecofeminism*, ed. Judith Plant [Philadelphia, PA: New Society, 1989], 1).

In other words, women have had to cultivate compassion in an unsustainable way, harming themselves and ultimately also others in the process. This destructive pattern should not be repeated now that compassion has become an important ecological virtue. It would be rather odd, and ultimately not effective either, if people would deplete their spirits in the name of ending the depletion of our soils.

7. See, for example, James D. Wallace, *Virtues and Vices* (Ithaca, NY: Cornell University Press, 1978), chap. 3; Alasdair MacIntyre, *After Virtue: A Study in Moral Theory*, 2nd ed. (Notre Dame, IN: University of Notre Dame Press, 1984), 192–3; Lee H. Yearley, *Mencius and Aquinas: Theories of Virtue and Conceptions of Courage* (Albany, NY: SUNY Press, 1990), 19. For a contrasting argument, see R. M. Hare, *Freedom and Reason* (Oxford: Oxford University Press, 1963), 149, 155, 187–91.
8. See, for example, Thomas Aquinas, *Summa Theologiae* II–II. 123.11, trans. the Fathers of the English Dominican Province (New York: Benziger Brothers, 1948).
9. The expression is ancient and can be traced to Plato's "Republic" (429 c-e) where courage is called a "preservative" (The "Republic," in *Five Great Dialogues*, trans. Benjamin Jowett [Roslyn, NY: Classics Club, Walter J. Black, 1942]); see also Thomas Aquinas, *Summa Theologiae* II–II. 123.12.
10. On "thick description" as a method of interpretation, see Clifford Geertz, *The Interpretation of Cultures* (New York: Basic Books, 1973), chap. 1; for an application of the notion to virtue theory, see Bernard Williams, *Ethics and the Limits of Philosophy* (Cambridge, MA: Harvard University Press, 1985), 129, 140, 143–5, 163, 192, 200.
11. See, for example, Alice Miller, *The Drama of the Gifted Child*, trans. R. Ward (New York: Basic Books, 1981), 64–75.
12. On grandiosity, see ibid., 38–42.
13. This dynamic bears resemblance to Aristotle's description of the proud man who "is the sort of man to confer benefits, but he is ashamed of receiving them; for the one is the mark of a superior, the other of an inferior" (*Nicomachean Ethics*, trans. W. D. Ross [Oxford: Oxford University Press, 1925], 1124^{b}5–10).
14. Complaints about human arrogance toward nature are widespread in ecoliterature. Again, Lynn White's essay, "The Historical Roots of Our Ecologic Crisis" set the tone by implicating "orthodox Christian arrogance toward nature" (*Science* 155 [10 March 1967], 1207). See also David Ehrenfeld, *The Arrogance of Humanism* (New York: Oxford University Press, 1978).
15. This strategy also does not generate genuine courage, but rather a semblance of it. For a similar argument, see, for example, Aristotle on rashness (*Nicomachean Ethics*, 1115^{b}25–35). More generally, Romanus Cessario observes: "Theories which suppose the will's supremacy over the emotional life treat the emotions as if they were unruly slaves in need of despotic control. Such repression of emotion, thought [sic] it may seem virtuous to some, amounts in effect to a miming of virtue" (*The Moral Virtues and Theological Ethics* [Notre Dame, IN: University of Notre Dame Press, 1991], 91).
16. Carolyn Merchant, *The Death of Nature: Women, Ecology and the Scientific Revolution* (San Francisco: Harper & Row, 1980), 223. Merchant draws on the work of Lynn White to make this point.
17. See, for example, Elizabeth Dodson Gray, "A Critique of Dominion Theology," in *For Creation's Sake: Preaching, Ecology, and Justice*, ed. Dieter Hessel (Philadelphia, PA: Geneva Press, 1985), 79–80.

18. For an alternative view on constituent attitudes of courage, see Edward Farley, *Good and Evil: Interpreting a Human Condition* (Minneapolis, MN: Fortress Press, 1990), 146–50. Farley argues that courage is "a complex posture that combines attitudes of relativizing all mundane goods at hand, consenting to the essential goodness and tragic character of the world, and venturing one's being in the perilous environments of the world" (150).
19. Aristotle, *Nicomachean Ethics*, 1117^{b}5–10.
20. See John Casey, *Pagan Virtue: An Essay in Ethics* (Oxford: Clarendon Press, 1990), 72. By contrast, ecofeminist Dorothy Dinnerstein associates so-called noble deeds with Thanatos, the human drive to destroy ("Survival on Earth: The Meaning of Feminism," in *Healing the Wounds*, ed. Plant, 196–7).
21. Thomas E. Hill, "Ideals of Human Excellence and Preserving Natural Environments," *Environmental Ethics* 5 (Fall 1983), 221.
22. Thomas Berry, "Economics: Its Effect on the Life Systems of This World," in *Thomas Berry and the New Cosmology*, eds. Anne Lonergan and Caroline Richards (Mystic, CT: 23rd Publications, 1987), 18. The other dimension of earthiness, i.e., down-to-earthness, is also stressed in the literature (see Chapter 2).
23. See, for example, Rosemary Radford Ruether, *Gaia and God: An Ecofeminist Theology of Earth Healing* (San Francisco: HarperCollins, 1992), 48, 251.
24. Roger S. Gottlieb, Introduction, in *This Sacred Earth: Religion, Nature, Environment*, ed. Roger S. Gottlieb (New York: Routledge, 1996), 4.
25. This claim also follows the more general belief that, in the words of Janis Birkeland: "If we want to get to the bottom of a psychological problem, we must uncover our 'blind spot,' or what we are denying." Birkeland warns, however, that "fear . . . should be understood also as a tool and product of manipulation." This requires social analysis, not only psychological work ("Ecofeminism: Linking Theory and Practice," in *Ecofeminism: Women, Animals, Nature*, ed. Gaard, 32, 36).
26. Challenging the traditional preoccupation with nobility need not raise the fear that we are opening a Pandora's box of other challenges. After all, the challenge is guided by specific reasons that apply only to this issue. At the same time we can continue to acknowledge our indebtedness to other aspects of the tradition.
27. Robert C. Roberts, referring to virtues of willpower (which include courage) as "capacities for self-management," suggests various techniques for dealing with emotions, such as deliberately behaving in ways that counter an emotion, and "skillful self-talk" ("Will Power and the Virtues," *The Philosophical Review* 93 [April 1984], 245). Roberts also points out, however, that these techniques should be complemented by a certain degree of passion and self-confidence in order to generate true virtue. Taking Roberts's argument one step further, I would argue that in the absence of a degree of passion and self-confidence, the use of these management techniques to deal with fear turns into a regime of heavy-handed control.
28. The situation is similar but more complex when a person experiences overwhelming fears that are pathological in origin. In such a case the person will have to begin by trying to understand and deal with the causes behind his or her fears. Therapy will have to precede channeling if harmful side effects are to be avoided. This process is particularly important because the control of pathological fears is not only likely to issue in undue attempts to control the environment, but also in subtle forms of abuse (for example, I once witnessed a rather Stoic friend of mine, marred by an inner-city childhood, purposely step on some crocuses that were gingerly peeking through the snow). Even in pathological

cases, then, the answer lies not in heavy-handed control (except perhaps temporarily as part of the therapy), but in a combination of therapy and imaginative channeling.

29. Wallace, *Virtues and Vices*, 76.
30. Aristotle, *Nicomachean Ethics*, 1115ᵇ20–30; Aquinas, *Summa Theologiae* II-II.126.1.
31. For a related argument, see Sharon D. Welch, *A Feminist Ethic of Risk* (Minneapolis, MN: Fortress Press, 1990). Welch argues that in a nuclear age we need to get beyond an ethic of control, which relies on chances of winning, to an ethic of risk, which involves "recognition that we cannot imagine how we will win" (19). An ethic of risk, in other words, involves genuine vulnerability.
32. Aristotle, *Nicomachean Ethics*, 1116ᵃ15–1117ᵃ30; Aquinas, *Summa Theologiae* II-II.123.1.
33. Josef Pieper, *The Four Cardinal Virtues* (Notre Dame, IN: University of Notre Dame Press, 1966), 138.
34. Ruether, *Gaia and God*, 268; Dorothee Soelle, *The Window of Vulnerability: A Political Spirituality* (Minneapolis, MN: Fortress Press, 1990).
35. A parallel argument concerning passions in general can be found in James D. Whitehead and Evelyn Eaton Whitehead, "Christians and Their Passions," *Warren Lecture Series in Catholic Studies* 21 (Tulsa, OK: University of Tulsa, 1992). As an alternative to the "mastery" of passions which, through denial, is likely to lead to "corrosive resentment or toxic sarcasm," the authors propose a process of "befriending": "Neither denying our emotion, nor abandoning ourselves to it, we acknowledge what we are feeling. We honor the arousal, confident it is not—finally—our enemy" (11).
36. While this proposal rests on the assumption that fear is an unpleasant emotion, and hence undesirable, it does not assume that "fear is bad" in the popular sense, that is, shameful.
37. Rosemary Ruether, for example, warns women not to express their environmental concern by repeating the romantic mistake of seeking to create a Garden of Eden in which men can find refuge (*New Woman/New Earth: Sexist Ideologies and Human Liberation* [New York: Seabury Press, 1975], 203).
38. Yearley, *Mencius and Aquinas*, 142–3; Yearley's observation is based on the *Summa Theologiae* I-II.69.3.
39. A similar transformation is described by Rosemary Ruether in "Courage as a Christian Virtue," *Cross Currents* 33 (Spring 1983), 8–16. Ruether argues that Christian courage is based on ultimate trust in God, expressed through faith, hope, and love. She contrasts "the courage of faith" to "the classic concept of courage, which involves militaristic self-mastery that was seen as only attainable by male aristocrats" (8–12).
40. I deliberately leave open the question whether the root of transformed courage is religious. Paul Tillich, at the end of his classic study *The Courage to Be*, takes a more definitive stance. Tillich argues that "every courage to be has an open or hidden religious root. For religion is the state of being grasped by the power of being-itself. In some cases the religious root is carefully covered, in others it is passionately denied; in some it is deeply hidden and in others superficially. But it is never completely absent. For everthing that is participates in being-itself, and everybody has some awareness of this participation, especially in the moments in which he experiences the threat of nonbeing" ([New Haven, CT: Yale University Press, 1952], 156).
41. I am reminded here of a line in the final act of Beethoven's opera Leonore: "True love knows no fear."

42. Augustine, *The Catholic and Manichaean Ways of Life*, trans. D. A. Gallagher and I. J. Gallagher (Washington, DC: The Catholic University of America Press,1966), 22.
43. Ibid., 34–5.
44. Ibid., 35.
45. See Yearley, *Mencius and Aquinas*: "religiously perfected courage becomes an inclinational not a preservative virtue" (14). On forms of benevolence, see Wallace, *Virtues and Vices*, chap. 5.
46. Robert C. Roberts also distinguishes between the ideal and the more normal situation: "The virtues of will-power are corrective in the significant sense that, in our present psychological condition but not in every imaginable one, they are needed to keep us on the path of virtue and our higher self-interest" ("Will Power and the Virtues," 233).
47. Ibid., 246, 247.

Chapter 9 Nine

NATURE AND ECOLOGICAL ETHICS: A WORKING TYPOLOGY

AT THIS POINT I can imagine that some of my readers may be wondering: All this talk about human tradition, human experience, and human discourse is nice and well, but should we not also consult *nonhuman* nature as a source for ecological virtue ethics? This is both a very important and a very tricky question, so I have kept it for the end. We should definitely pay attention to nonhuman nature, if only for the simple reason that it is authoritarian to pursue an ethic that affects stakeholders without involving these stakeholders in the process. The tricky point, however, is to figure out what "involvement" means in the case of nonhuman nature.

Already two decades ago, Holmes Rolston highlighted some of the difficulties involved in an article with the straightforward title, "Can and Ought We to Follow Nature?"[1] Rolston's typology of answers continues to be very helpful.[2] However, since his 1979 publication, the discourse of ecophilosophy and ecotheology has been much developed. Nonhuman nature occupies a great variety of roles in this discourse, and the theme of "following nature," which inspired Rolston's typology, cannot fully capture all these roles. Moreover, nonhuman nature is represented by multiple images, so that in order to gain meaningful insight into the role of nature in ecological ethics one wants to ask, Which nature?

In response to these developments I have designed a two-dimensional typology, which I offer here both as a continuation of Holmes Rolston's pioneering work and as an invitation to extend the discussion more specifically to the area of virtue ethics. Following the interpretive character of this book, I first use the typology as a hermeneutical key to improve our understanding of the ways in which nature is rendered morally operative in existing eco-discourse. Thomas Berry's *The Dream of the Earth* and Murray Bookchin's *The Ecology of Freedom* again serve as focus texts. The different ways in which these authors let nature play a role in their ethics will help us appreciate the complexity of the issue. At the same time the precise analysis made possible by working with a typology will enable us to gain insight without being unnecessarily haunted by the ghost of the naturalistic fallacy.[3] The bulk of the chapter is devoted to this interpretive task. Keeping in mind the complexity and options presented by existing discourse, I then proceed on a more normative track with an outline of how, with the aid of the same typology, we can be creative, critical, rhetorical, and pluralistic when we involve nonhuman nature in building an ecological virtue ethic.

A Two-Dimensional Typology

The typology I propose is two-dimensional, which means that it can accommodate three variables: one on each axis and one in the field thus created. The typology not only lists the variables, but also shows how they are interrelated (see Appendix B). On the horizontal axis I list seven *functions* that typify key interests and concerns of Western ethical discourse—hence the term "typology." For the sake of clarity, I will present them as analogous to various aspects of playing chess.[4] These functions are: (1) "impetus," which refers to the basic motivation to be moral (cf. what makes two people decide to play a game of chess); (2) "support," which refers to the preconditions that must be met for character formation, moral reasoning, and action to be possible (cf. the food, drink, and sleep that keeps the chess players alive and alert); (3) "limitation," which refers to the boundaries of permissible/helpful human action (cf. knowing the limits of the chessboard and the prohibitive rules of the game); (4) "direction," which refers to the need for general purposes and guidelines (cf. a master chess strategy); (5) "preservation," which refers to the need to know when and how general interests must be protected in concrete situations (cf. defensives moves); (6) "construction," which refers to the need to know how general aims can be approached through concrete steps (cf. knowing how to advance); and (7) "choice," which refers to the necessity to prioritize in situations of conflicting needs (cf. knowing when to sacrifice a chess piece in order to maintain one's chances of winning). Since these seven functions reflect ethical concerns that go beyond virtue discourse, the typology has a more general applicability than to ecological virtue ethics only.

The vertical axis contains *images* of nature that occur either explicitly or implicitly as metaphors in ecodiscourse. The sample typology in Appendix B includes the images of divine power, cosmic embrace, book of law, matrix, pyramid, community, and respected other. This list is not intended to be exhaustive; it merely reflects the images discussed in this chapter. I invite my readers to add images to the list and to see how they correlate with the various ethical functions, either in actual ecodiscourse or as a constructive exercise. The images I do mention are arranged on a sliding scale, according to different levels of visual perception. I start with macroimages, by means of which we try to capture cosmic dimensions (e.g., divine power, book of law). Mesoimages most strongly connote planetary dimensions (e.g., matrix, pyramid). Microimages particularly suggest bioregional dimensions, including the level of individual entities (e.g., community, respected other).[5] By listing images—that is, metaphors—rather than concepts, I want to draw attention to the role of visual imagination in the ethical discourse in which nonhuman nature plays a role. Ecodiscourse contains far more images of nature than rationally developed concepts, and, as I will illustrate, these images greatly influence which, if any, moral roles authors are willing to assign to nature. I also believe that there are advantages to working with metaphors in ecological ethics. These advantages I will discuss at the end of the chapter.

In the typological field that represents the correlation between ethical functions and images, I locate the concrete moral *roles* that nature occupies in ecodiscourse. In other words, here we find answers to the question: "How does nature, represented as Y, fulfill ethical function X?" Thus roles are specific ways of fulfilling a function, just as the roles of a teacher or a father involve specific ways in which children can be educated.

In the next two sections I illustrate how a two-dimensional typology of ethical functions, images, and moral roles can be used as a hermeneutical key to improve our understanding of the ways in which nonhuman nature enters moral reflection in Thomas Berry's *The Dream of the Earth* and Murray Bookchin's *The Ecology of Freedom*. I discuss Berry's work first, because his strong holism assigns certain roles to nature that Bookchin, from a position of weak holism, finds unacceptable.[6] Berry's thought provides a background that puts Bookchin's reservations in perspective. The results of this interpretive journey are included in the sample typology in Appendix B.

NATURE AND ETHICS IN *THE DREAM OF THE EARTH*

As a hermeneutical key, a two-dimensional typology of ethical functions, images, and moral roles of nature can accommodate multiple interpretive questions. In this section I will illustrate how the typology can aid in interpreting what I call morally laden nature passages, that is, passages in which nature somehow figures as the source of, or as contributing to, a moral "ought." By means of the typology, we can clearly sort out which moral roles are attributed to nature, from which metaphor(s) these roles derive their meaning and power, and to which ethical function(s) they relate. I will examine four passages from *The Dream of the Earth* that together address the entire range of ethical functions.

The first passage reflects several important themes in Berry's work. The focus is on the universe as an expanding process, the history and direction of which are sedimented, as it were, in the genetic structure of the universe's most complete mode of self-expression—the human being. The intricate way in which universe and humans are related clearly has moral implications for Berry:

> In moments of confusion such as the present, we are not left simply to our rational contrivances. We are supported by the ultimate powers of the universe as they make themselves present to us through the spontaneities within our own beings. We need only become sensitized to these spontaneities, not with a naive simplicity, but with critical appreciation. This intimacy with our genetic endowment, and through this endowment with the larger cosmic process, is not primarily the role of the philosopher, priest, prophet, or professor. It is the role of the shamanic personality. (*DE* 211)

In this passage the universe provides clarity to confused humans by means of a process reminiscent of religious revelation, a process mediated through human genetic structures.

Using the categories of the typology, we can say that: (1) nature is portrayed by means of language that connotes omnipotent personal agency ("powers that make themselves present to us"), suggesting the operation of traditional religious *imagery* such as divine power or world spirit;[7] (2) befitting this imagery, nature is cast in the *role* of a revelatory agent that provides clarity "in moments of confusion"; (3) as such, nature especially addresses the ethical *functions* of: (a) "impetus," because the manifestation of a good, trustworthy, authoritative power experientially solves the question of "why be moral" (for those, that is, who are inclined to submit themselves to a larger power in the first place); (b) "direction," because the confusion that nature removes can be a matter of needing moral vision; (c) "construction," because confusion can also be a matter of needing a concrete plan of action; and (d) "choice," because confusion can consist in not knowing which of various avenues to pursue.

The next passage shows an entirely different picture and role of nature in connection with two of the same ethical functions addressed by the previous passage. Thus we get a sense of the variety of ways in which nature becomes morally operative, even in the work of a single author:

> We might consider our intimate and compassionate presence to the earth as originating ultimately in the curvature of space, as it is presented in modern science. The entire earth community is infolded in this compassionate curve whereby the universe bends inwardly in a manner sufficiently closed to hold all things together and yet remains sufficiently open so that compassion does not confine, but fosters, the creative process. (*DE* 20)

In this passage the human virtue of compassion toward the Earth appears to be the expression of an overarching cosmic design. To be more precise, one could say that: (1) nature is portrayed as a "compassionate curve"—a cosmic embrace as it were—an *image* inspired by modern science; (2) as such, nature is cast in the *role* of structuring the human capacity for a particular virtue, a role commonly found in ancient Asian traditions;[8] (3) in this role, nature addresses the ethical *functions* of "direction" as well as "construction," because virtues express general goals (sometimes referred to as "excellences" or as "flourishing") but these goals are only recognizable through the concrete actions of virtuous persons.

A third example puts nature in a role more familiar to Western ethicists. Leading up to the following passage, Berry argues that the evolutionary process of the universe has been controlled by "governing principles." Their moral relevance as well as their content becomes clear when he writes:

> The ecological age must now activate these principles in a universal context if the human venture is to continue. These principles on which the universe functions are three: differentiation, subjectivity [a "capacity for interiority" associated with "numinous quality"], and communion. (*DE* 45)

In other words, Berry translates natural principles of teleomatic development into moral laws. For example, elsewhere in *The Dream of the Earth* he describes the "law of communion" as stating that "every reality of the universe is intimately present to every other reality of the universe and finds its fulfillment in this mutual presence" (*DE* 106, 107; this definition derives its normative character from the term "fulfillment," which is a teleological concept).[9] In terms of the categories of the typology, we can say that: (1) the above passage suggests the *image* of nature as a book of law, a metaphor with ancient Stoic as well as Christian roots;[10] (2) portrayed as such, nature is cast in the authoritative *role* of legislating morality; (3) by fulfilling this role, nature addresses the ethical *functions* of "direction" as well as "limitation," since moral laws both provide general guidelines and suggest negative repercussions for noncompliance.

In a final example of a morally laden nature passage, too long to quote entirely, Berry focuses on the bioregion as the "context [which] not only activates our interior faculties . . . [but] also provides our physical nourishment" (*DE* 165). As a tightly knit "community," the bioregion requires of its members—if they and the community are to survive—that they integrate their functioning within the functioning of the larger whole. Thus, humans must model their actions on what Berry identifies as six "community functions" of the bioregion. I will briefly discuss each function, since here we find a very detailed code of ecological ethics modeled directly on the natural world (cf. *DE* 166–8).

First of all, the bioregional function of self-propagation "requires that we recognize the rights of each species to its habitat, to its migratory routes, to its place in the community" (*DE* 166). Second, the function of self-nourishment mandates that humans work within patterns of mutual nourishment and that there be no waste. Third, the function of self-education suggests that humans must avail themselves of "the instruction available through the natural world" (*DE* 167). Fourth, the function of bioregional self-governance operates through an order of interior bonding in which humans must integrate themselves. Fifth, the function of self-healing suggests that humans, in their attempts to seek healing, submit themselves "to the discipline of the community and acceptance of its nourishing and healing powers" (*DE* 168). And sixth, the function of self-fulfillment suggests that, just as a field finds fulfillment in its flowering, humans must pursue their own proper mode of fulfillment, which, according to Berry, lies especially in "conscious celebration of the numinous mystery of the universe expressed in the unique qualities of each regional community" (*DE* 168).

By means of the categories of the typology, we can highlight just how comprehensive an ethic is presented in this passage: (1) portraying nature at a bioregional level, Berry uses the *image* of a community, an image that already connotes virtues such as fitting-in and contributing; (2) the passage casts nature in four distinct moral *roles*: (a) the community "activates our interior faculties" and "provides our physical nourishment," thus creating the conditions for human moral agency (i.e., an enabling role); (b) it demonstrates the limits of survival by showing what goes wrong if individuals and species do not work within the community (i.e., a warning role); (c) consequently, it mandates integrated functioning; and (d) the community provides models for such integrated functioning;[11] (3) with these four moral roles the passage addresses five ethical *functions*: The first three roles correlate to the functions of "support," "limitation," and "direction," while the last role correlates to both "preservation" (the community models effective ways of propagation, nourishment, healing, and governance, all of which contribute to preservation) and "construction" (the community provides a model for education and self-fulfillment, both of which involve taking concrete innovative steps).

By using the typology as a hermeneutical key, we can highlight images, moral roles, and ethical functions of nature in morally laden nature passages. In the four passages I have used as examples, nature emerges variously as a divine power, a cosmic curve, a book of law, and a community. It is cast in authority-providing, revealing, structuring, legislating, enabling, warning, mandating, and modeling roles. And it addresses all the ethical functions of impetus, support, limitation, direction, preservation, construction, and choice. Thus, with the help of the typology we can begin to appreciate the variety of ways in which nature becomes morally operative in *The Dream of the Earth*, even by considering only a few passages.

By using the same hermeneutical key to interpret all the morally laden nature passages in an entire book, or even in all the works of a particular author, we will be able to infer patterns. For example, we may find that a writer thinks of nature primarily in macroimages, and that these images are more conducive to inspiring certain roles and functions than others. Or we may discover that some ethical functions are not addressed at all in a particular work, while others are developed in great detail through rich clusters of roles and metaphors. Such findings will make an important contribution to our understanding of the role of nature in current ecodiscourse. This will in turn, enable us to be more precise in our constructive efforts.

NATURE AND ETHICS IN *THE ECOLOGY OF FREEDOM*

Unraveling morally laden nature passages is only one hermeneutical endeavor facilitated by the typology. We can also employ the categories of the typology for more directed searches. For example, we can investigate if a particular moral role occurs in the work of an author, and if so, how it correlates with metaphors of nature and ethical functions. Or we can start with a key metaphor and investigate which, if any, moral roles it inspires and which ethical functions it addresses. I have selected a passage from Murray Bookchin's *The Ecology of Freedom* to illustrate each of these hermeneutical approaches. This exercise in interpretation brings to light some striking differences between Berry and Bookchin when it comes to the role of nature in ethics.

Say we want to find out if nature functions in the role of a moral model in *The Ecology of Freedom*, like it does in Berry's work. The following passage suggests a strongly negative answer, while indicating a cluster of metaphors and roles that do seem important to Bookchin:

> I cannot emphasize too strongly that *nature itself is not an ethics*; it is the *matrix* for an ethics, the source of ethical meaning that can be rooted in objective reality. Hence nature, even as the matrix and source of ethical meaning, does not have to assume such delightfully human attributes as kindness, virtue, goodness and gentleness; nature need merely be *fecund* and *creative*—a *source* rather than a "paradigm."
>
> The function of an ethical philosophy does not entail a mimetic reduction of ethics to its source. Rather, it requires a *ground* from which to creatively develop ethical ideals. (*EF* 278)

For example, Bookchin explains:

> to transmute "pacification" [e.g., the image of the wolf lying down with the lamb] into "domestication" is to deal with nature as a model of ethical

> behavior rather than to accept it for what it really is—a source of ethical meaning that reestablishes our sense of ecological *wholeness*, the underlying dialectic of unity in diversity. (*EF* 278)

According to Bookchin, nonhuman nature cannot be a model for ethics because, even though humans are related to nonhuman nature, they are still significantly different. People who overlook or deny this difference tend to project human characteristics onto nonhuman nature.[12] When such projections are elevated to the role of moral model, we find a circular and self-deceptive form of moral reasoning. Thus Bookchin deliberately rejects one particular moral role for nature that plays a central role in Berry's work.

Even so, this passage from *The Ecology of Freedom* suggests that Bookchin does see a place for nature in ethics. The key to understanding Bookchin's position on this point seems to be a cluster of metaphors he uses interchangeably, suggesting a single *image*: nature as matrix/source/ground. How does this image become morally operative? The matrix/source/ground, Bookchin claims, provides "ethical meaning," which "reestablishes our sense of ecological *wholeness*" and helps us "creatively develop ethical ideals." These three phrases, when interpreted in light of the typology, suggest three roles, which can be related to three ethical functions.

When Bookchin states that nature, as matrix, provides "ethical meaning," he has in mind the fact that we perceive in nature a consistent, that is, historical development:

> The matrix from which objective reason may yet derive its ethics for a balanced and harmonized world is the nature conceived by a radical social ecology—a nature that is interpreted nonhierarchically, in terms of unity in diversity and spontaneity. Here, nature is conceived not merely as a constellation of ecosystems but also as a meaningful natural *history*, a developing, creative, and fecund nature that yields an increasing complexity of forms and interrelationships. (*EF* 274)

Bookchin assumes that this evolutionary process is a good, developing in a right direction, which is why it takes on *ethical* significance for him. In terms of the typology, we can say that providing ethical meaning is a moral *role* which Bookchin assigns to nature.[13] This role fulfills the *function* of furnishing an impetus for the moral life, a reason to be moral: Since we can trust that there is something good going on, it makes sense to contribute our best efforts.

By contrast, the statement that nature, as a source, "reestablishes our sense of ecological *wholeness*," suggests a moral *role* that fulfills the ethical *function* of providing support for the moral life. *The Ecology of Freedom* gives a detailed analysis of the historical emergence of repressive social structures that alienate human beings from their innate ability to perceive unity in diversity, and also from their creativity and spontaneity. Contact with nature can stimulate restoration of these capacities, which are essential for the moral life. There is nothing mystical about this for Bookchin. We are not dealing with a numinous source of energy, a personified cosmic healing agency. Rather, as a source nature is simply the context of dynamic relationships that triggers a readjustment of our perspective to what it would have been anyway if misguided cultural forces had not interfered.[14] It supports the moral life like a family vacation can restore the health and perspective of a stressed-out workaholic.

The statement that nature helps us "creatively develop ethical ideals" (*EF* 278) suggests yet a third moral role associated with the matrix image, a role that fulfills the ethical function of providing general direction to the moral life. For example, ethical ideals like freedom and self-consciousness play prominent roles in *The Ecology of Freedom*, and they are directly inspired by nature as a source of ethical meaning. Bookchin perceives a consistent development toward ever greater variety and subjectivity in nature, and human freedom and self-consciousness are culminations of this development.[15] Thus nature as matrix not only provides the moral life with its raison d'être and the necessary healing support, but it also helps provide ethical content: a sense of direction defined by clear ideals.

These examples from *The Ecology of Freedom* show how even relatively simple directed searches guided by the categories of the typology (starting, respectively, with a modeling role, and then with the image of matrix/source/ground) can yield a detailed articulation of the way in which nature becomes morally operative in a particular work. The examples also show how the typology can facilitate comparative research. From the few passages discussed, we can already learn that nature takes on a modeling role for Berry but not for Bookchin, and that Berry sees nature as authoritatively shaping and guiding human moral agency, whereas Bookchin understands human virtues such as ecological sensibility, creativity, and spontaneity not in any way as steered by nature, but rather as revived and enriched by it. Although one should not conclude on the basis of these few passages that Berry and Bookchin hold completely incompatible views on the place of nature in ethics, these are significant differences. By facilitating the clear articulation of such differences, the typology can contribute to the search for appropriate and helpful ways in which nature can be morally operative in ecological ethics.

Normative Uses of the Typology

I have shown how, as a hermeneutical key, a two-dimensional typology of ethical functions, images, and moral roles of nature can accommodate multiple interpretive questions (Which image underlies this role? How does nature address a particular ethical function in this work? Which moral roles and functions are linked to a particular image?, etc.). The typology can also help us be precise and consistent as we take a more normative approach and articulate our own views on the role of nature in ecological ethics, including virtue ethics. In the remainder of this chapter I will outline four complementary options.

First, in addition to serving as a *hermeneutical key*, the typology can also serve as a *heuristic key* by means of which we can systematically generate and try out constructive ideas for involving nature in our ethical theories. The open spaces in the typological field (see Appendix B) can be taken as invitations to ask, "Could image Y also fulfill function X, and how exactly would this work (x, y)?" Or, "Could function X be addressed by image Y and what kind of role (x, y) would this most likely imply?" Particularly pertinent is the question of which images could be helpful in addressing the ethical function of "choice," since here we find some of the most pressing practical problems. Using the typology we could hypothesize, for example, that the task of setting priorities in situations of conflicting needs requires a *balance* of at least two images: one that facilitates making decisions that unavoidably require sacrifice (e.g., nature as pyramid, which suggests species ranking as a way to make choices), and one that prevents the process from degenerating into cold and purely instrumental calculation (e.g., nature as respected other, which suggests intrinsic value throughout much or all of the natural world).[16] This example illustrates how the typology can be used to stimulate systematic constructive reflection. By creatively filling the open spaces in the typology (like a crossword puzzle), we generate ideas that contribute to the search for logically sound and psychologically wise involvement of nature imagery in our ethics. These ideas have great practical value at a time when new problems constantly call for new avenues of moral reasoning.

Second, the typology can serve as a *critical key* to facilitate analysis of the ideological roles of nature imagery in ecodiscourse.[17] Some images, like "mother" and "pyramid," have already received extensive critical attention. Yet other images are widely assigned moral roles without much critical awareness. I think, for example, of the metaphors "kin" and "web," the use of which seems virtually obligatory in some ecodiscourse, a signal that they may be part of a new power play. Images with strong holistic overtones, such as "organism" and all personifications of the Earth or the Universe, also require special critical alertness: If these images are regularly connected to roles of authority (e.g., calls to "obey," "fit in," "follow"), we should ask who is setting up this authority and with what motives.

Third, the typology can also be used as a *rhetorical key* to help identify which images of nature would be most effective in making a particular ethical function operational. For example, it would be unwise to draw on the image of nature as a book of law when modern Western people, many of whom rebel against top-down legalism, need to be energized to protect a rain forest (preservation) or to start neighborhood cooperatives for production and exchange (construction). Images of nature as a web or community would more likely have popular appeal in these cases.[18] Overall, discernment of appropriate rhetorical images is crucial for fruitful moral agency, and the typology here presented can sharpen such discernment.

Finally, the typology can serve as a *pluralistic key* to aid in the development of nonarbitrary, pluralistic forms of ecological ethics. If we want to honor and encourage *multicultural* moral empowerment, it is important to allow for and develop *multiple* metaphors of nature in the discourse of ecological ethics. The typology, by openly listing many images of nature, contributes to the acceptance of such variety and to mutual trust, while ensuring that real differences are not pushed under the carpet. Let me illustrate this point.

First of all, clearly not every image of nature is acceptable to everyone, but the typology shows that we need not immediately fear that the moral agency of some persons or groups who do not accept our favorite image is therefore necessarily deficient in certain functions. It is very unlikely that only one image would be able to address a specific ethical function. This will become obvious as soon as more images are added to the sample typology in Appendix B. In its current version the function of "direction" is already addressed by five images: divine power, cosmic embrace, book of law, matrix, and community. Therefore, if a person or group cannot accept one or several of these images, this does not necessarily imply that their moral agency will lack general direction.[19] Thus the typology helps to maintain trust in the presence of metaphorical difference.

This even applies to situations in which, because of fundamental metaphorical differences, people may be inspired to engage in mutually incompatible actions. (That need not always happen, of course; different metaphors may also inspire similar moral actions.) In such potentially conflictual cases, the typology serves as a reminder that at least no party has been robbed of the nature imagery that fits their cultural heritage and personal history and that thereby has the power to inspire. The last thing we need in an ecological age is the rule of *one* metaphor of nature, a metaphor that in the interest of ending human domination of nature would dominate human moral imagination—which would surely soon bring it to an end.

This takes me to my final observation on the use of a two-dimensional typology based on metaphorical representations of nature. Metaphors derive their power from a tension between similarity and difference. Someone who calls nature a book of law, for example, highlights certain features of nature that remind him or her of law, such as regularity and consistency. At the same time this person knows that nature is not a real book with printed pages in which one can read laws. Thus a nature metaphor says as much that nature *is not* something as it says that nature *is* something.[20] This point is underscored even further when a multiplicity of metaphors is used, for clearly nature could not "really" be so many different things at once.

In a sense, we are confronted here with a failure of human language. Nature constantly escapes our attempts to describe and hence understand it. Yet this failure contains a blessing in disguise. Awareness of the metaphoric character of images of nature ultimately lets nature be free.[21] To the extent that the development of a typology of metaphors contributes to this awareness, it helps the development of a genuinely liberating ecological ethic.

NOTES

1. Holmes Rolston III, "Can and Ought We to Follow Nature?" *Environmental Ethics* 1 (1979), 7–30.
2. In his typology, Rolston distinguishes between following nature in (1) an "absolute sense" (we cannot but follow nature, since we are nature); (2) an "artifactual sense" (we cannot follow nature, since human action alters the course of nature); (3) a "homeostatic sense" (we can use the laws of nature as instruments to serve our own well-being); (4) an "imitative ethical sense" (we can imitate nature, but this is not morally recommendable, for nature is too cruel); (5) an "axiological sense" (we are drawn to preserve and contemplate the intrinsic value we encounter in nature); and (6) a "tutorial sense" (we experience nature as a leading force and are shaped by its wisdom). In this progression of types, Rolston moves from logical analysis to an articulation of his own moral intuitions.
3. With the expression, "naturalistic fallacy," G. E. Moore indicated what he saw as the mistake of moving from a natural *is* to a moral *ought* (*Principia Ethica* [Cambridge: Cambridge University Press, 1903, 1956], 36–58, 188, 193, 195–8, 200, 206).
4. Like all analogies, the comparison between ethics and chess breaks down at a certain point. Personally, I do not like to think of ethics as a "competitive" endeavor.
5. When images apply to multiple levels, they can be placed in the middle of their range. Neither the functions nor the images listed in Appendix B are arranged in such a way as to lead up to a "preferred type" (as one so often finds in typologies). My personal preference goes to images that reflect a weak type of holism.
6. I use the term "strong holism" to describe the idea of a single substance permeating the universe, and "weak holism" to describe the idea of the universe as an integrated system. For further discussion, see chapter 2.

   References in the text will be to Thomas Berry, *The Dream of the Earth* (San Francisco: Sierra Club Books, 1988), referred to as *DE* and to Murray Bookchin, *The Ecology of Freedom: The Emergence and Dissolution of Hierarchy* (Palo Alto, CA: Cheshire Books, 1982), referred to as *EF*.
7. Here one notices the influence of Christian (immanentist) spirituality in Berry's work.

8. Berry observes, for example: "What is remarkable thoughout the Asian world is that terms designating supremely affectionate qualities carry ultimate cosmological significance. So in the Chinese world, *jen*, a term translated as love, benevolence, or affection, is not only an emotional-moral term, it is also a cosmic force. This can be said also of the virtue of *ch'eng*, translated as sincerity or integrity. In India the term *bhakti*, devotional love, was a cosmological as well as a spiritual force. In Buddhist tradition the term *karuna*, compassion, is a supreme cosmic power. Thus we find a pervasive intimacy and compassionate quality in the very structure of the universe and of the earth itself" (*DE* 20).
9. See also: "It is the mission of our present and all future generations to develop this capacity for mutual presence on new and more comprehensive levels" (*DE* 135).
10. Berry draws an explicit link with medieval natural law tradition. However, he points out that modern people cannot assume the fixed universe that informed natural law tradition until recently (*DE* 105).

    Interesting in this context is that Berry alternates the terms "law" and "capacity". Paralleling Thomas Aquinas's natural law theory, he seems to assume that natural laws are matched by human capacities. See: "We have not adequately developed this capacity for communion. We have been especially delinquent in fulfilling this law of communion in relation to the natural world" (*DE* 107).
11. Elsewhere Berry also presents the Earth as a model of an evocatory rather than a dominating attitude (*DE* 14), and as a model to the professions (*DE* 68, 107).
12. In addition to the projections cited in this passage, Bookchin elsewhere also criticizes the idea that nature is passive and feminine (*EF* 33), or stingy and cruel (*EF* 276–8).
13. Bookchin is aware, though, that this is a matter of interpretation: Nature simply provides enough consistent development to stimulate humans to discern moral significance. Thus Bookchin does not simply "derive value from facts"; he does not commit the naturalistic fallacy.
14. It is important to point out that even in this instance, nature does not function as a model. For example, elsewhere Bookchin stresses the difference between human and natural spontaneity: human spontaneity must be supplemented by will, insight, and knowledge, whereas "in nature . . . spontaneity operates within a more restrictive set of conditions" (*EF* 36). Thus the restoration of human spontaneity through contact with nature does not rest on simple imitation.
15. See chapter 3 on the teleological character of Bookchin's theory.
16. This example metaphorically expresses Lawrence Johnson's argument that we need to distinguish between two complementary ideas: "moral significance," which helps in making choices by assigning greater significance to entities with greater complexity, and "moral consideration," which is used to ascribe inherent value. Both are necessary. (See *Morally Deep World: An Essay on Moral Significance and Environmental Ethics* [Cambridge: Cambridge University Press, 1991]).
17. I am indebted to Hans Dirk van Hoogstraten (Department of Theology, Catholic University of Nijmegen, the Netherlands) for raising the issue of ideology criticism in this context.
18. I do not want to suggest, however, that the use of metaphors should be entirely instrumental. A "use whatever works" kind of sophistry would be superficial and could degenerate into dangerous demagoguery. The method of discernment I propose is only intended to make wise choices between metaphors that already have a basis in a people's historically conditioned experience. I do not

propose that, for the sake of moral efficacy, people should switch to (or worse; be forced to switch to) metaphors with which they have no cultural connection.

19. In general it seems that the range of images that can be morally operative in an ethic inspired by weak holism is smaller *in principle* than the range that can be used in an ethic inspired by strong holism, simply because most weakly holistic images are consistent with strong holism, whereas the reverse is not true. This does not mean, however, that weak holism therefore necessarily leads to a less diverse ethic in practice.
20. On this quality of metaphors in general, see Sallie McFague, *Models of God: Theology for an Ecological, Nuclear Age* (Philadelphia, PA: Fortress Press, 1987), 33.
21. Catriona Sandilands makes a similar point with regard to nature language in general: "Any language of nature is destined to fail, and this failure requires recognition of a moment of nature that overflows any attempt to capture its positivity" ("From Natural Identity to Radical Democracy," *Environmental Ethics* 17 [1995], 88).

CONCLUSION

THE BIGGEST DANGER THAT threatens books on virtue is death by analysis. Life-giving habits and attitudes that flourish in an atmosphere of spontaneity have a proclivity to wither when subjected to the light of critical inquiry. This is an old problem. Generations of Christians have been baffled by the paradoxical task of cultivating humility while not being too aware of their successes, lest they become proud or too preoccupied with the rewards that would await them in heaven. Similarly, ecological virtues like respect for nature, attunement, frugality, compassion, courage, and care are ideally cultivated without too much self-awareness and theoretical hairsplitting. Thus they will more likely involve authentic self-expression, uncalculated self-giving, and intuitive sensitivity to the situation at hand. As soon as they become objects of reflection, ecological virtues risk becoming part of some new "program" that people "ought" to follow, a program that introduces external rewards in the form of social acceptance for political correctness. Such a development not only spells the end of these virtues as virtues, carrying their own rewards, but also of their cultivators' unambiguous orientation toward nonhuman nature.

Perhaps the authors I have studied have intuitively understood this and have wisely chosen not to make a big theoretical deal about the virtues they espouse. Perhaps I have broken a taboo—the kind that has a good reason to exist—by writing this book. If this is so, I ask for forgiveness and hope that what I have written will again be forgotten.

But, like most sinners, I also want a chance to explain my deed. When I first discovered ecological virtue language, this discovery struck me as extremely good news. Here was a moral language that expresses commitment and hope in bleak times; a language that combines reason and emotion, spontaneity and discipline, self-development and self-sacrifice, all in holistic ways consistent with ecologically informed worldviews; and a language that has roots in traditions with demonstrated adaptability to new circumstances (an important feature given the unprecedented size and character of many current ecological problems). Best of all, this wonderfully appropriate language was directed not simply at human happiness but at the good of the entire Earth community! Such good news had to be communicated.

Of course communication led to the temptation to say a few words about cultivation also. The reader who has worked through Part Two may be under the impression that an ecologically virtuous life is not manna fallen from heaven but bread that has to be earned in sweat and toil. I hope that my reflections on the use of moral sources and on criteria to identify vices and cultivate genuine virtues will not dampen anyone's moral enthusiasm but rather stimulate creative and critical reactions. The good news that a rich and vibrant and promising ecological virtue discourse exists requires our thoughtful attention. Only if carefully nurtured will this discourse avoid pitfalls that would undermine its vision of a flourishing Earth, pitfalls buried in older habits and social structures that continue to corrupt even our best intentions. I have gambled on there being a middle ground between death by analysis and death by lack of critical awareness.

APPENDIX A: A CATALOGUE OF ECOLOGICAL VIRTUES AND VICES

The following 189 virtues and 174 vices occur in post-1970 environmental literature. Synonyms are listed (and counted) together.

VIRTUES

acceptance of limitations/mortality
accountability
adaptability/flexibility
adoration (of Creator and creation)
advocacy
affirmation (of others)
agape/altruism
anger/resentment
appreciation/recognition
asceticism
attentiveness
attunement
autonomy/disengagement/independence (from commercial life)
awe/fear
being informed/knowledgeable
beneficence/active goodwill
benevolence
biocentrism/life-centeredness
bioregionalism/local awareness/ regionalism
breathing (with awareness)
calmness
care
carefulness/cautiousness
celebration
challenging
channeling emotions
clarity (of vision)
commitment
communication
compassion
competence
complementarity
concern
confronting despair
conservation
constructiveness
contentment/sufficiency
cooperation
cosmocentrism
courage/heroism/daring
courtesy/graciousness
creativity/imagination
critical reflection/mistrust
curiosity
decency
determination
deep thinking/feeling
development/maximization (of creation)
diligence/industriousness
discernment/seeing/perception
discipline
dwelling-in-place/sense of place/patriotism
dwelling-in-the-moment/alertness
dwelling-with-wounds
earthiness
ecological sensibility/libertarian rationality
efficiency
empathy/sympathy
empowering
endurance/long-suffering

equanimity
eros/lust
eschatological attitude
evolutionary remembering
expressing feelings
faith/piety
forgiveness
fostering
friendliness
friendship
frugality/thrift
generosity
gentleness
getting in touch with feelings
grace/gracefulness/elegance/delicacy
gratitude
harmonizing with the will of the land/
flowing with Nature
healing
helping
holistic thinking
honesty/truthfulness
honorableness/honor/sense of moral
probity
hope
hospitality
humility
humor/wit/hilarity
identification
inclusivity
individuality/having personality
integrity/consistency
intellectual sophistication
interdependence
intimacy/communion
introspectiveness/inward spirituality
intuitiveness
invitational attitude/evocative attitude
joyfulness
justice/ecojustice
kindness
leadership (by example)
learning
letting be/live and let live
liberation
listening
living lightly on the Earth
long-range thinking
love/charity
loyalty/faithfulness/devotion/fidelity
managing
mastery/exercising dominion
maturity/emotional fortitude
meditation/reflection
meekness
mercy
modesty
mourning
naiveté (postcritical)
neighborliness
nonmaleficence/noninjury/*ahimsa*
nonviolence/pacifism
nurturing
openness/receptivity
organic knowledge
participation/communal sensibility
passion/being passionate/enthusiasm
patience
peacefulness/being quiet
peacemaking
perseverance/strength
poverty (voluntary)
praising
presence
preservation
protectiveness
prudence/good judgment
rage
realism
rebelliousness/liberatory sensibility/allied
resistance
reconciliation/conciliatory sensibility
relationality/sense of connectedness
relinquishment/abandonment/"dissipating"
reluctance to impose
repentance/confession
respect/honor
responsibility/responsiveness
reverence/other-esteem
sacrificing/cross-bearing/sacrificial love

self-confidence
self-defense
self-denial
self-examination/self-reflexivity/ self-criticism
self-love
self-realization
self-reliance
self-respect
self-understanding
sense of ecological self
sense of kinship
sense of ritual
sense of sacramentality
sensitivity
sensuousness/passion for the concrete
servanthood
sharing
simplicity (voluntary)
sincerity
sobriety
solidarity/sense of unity
spirituality
spontaneity
stewardship
submission/obedience/alignment
supporting
sustainability
temperance/moderation/restraint
tenderness
tending
theocentrism
theoretical discrimination
thoughtfulness/consideration/ mindfulness
tolerance/forbearance
trust/confidence
understanding
vigilance
vision/utopian thinking
vulnerability
warmth
wisdom
wonder

VICES

abstract thinking
abusiveness
aggressiveness
alarmism
alienation/dislocation (from one's body, from the earth)
ambition (in excess)
analytical thinking
androcentrism
anger/rage
anthropocentrism/anthropomonism
apathy
arrogance
asceticism
atavism
atomistic thinking
attachment (to wealth, power)
authoritarianism/authoritarian rationality
avoidance of death
blasphemy
blindness
brashness
bureaucratic attitudes
callousness
carelessness/lack of care
chauvinism
coerciveness
competitiveness
complacency
condescension
conformity
conquering attitude
consumerism/materialism
contempt/disdain (for physical realities)
control
corruption
cowardice
cruelty/sadism
cynicism
deceit
defensiveness
denial

dependence (on modern business corporations)
desecration (of nature)
despair
destructiveness/vandalism
discouragement
disobedience
disloyalty
distancing
distrust/suspicion
dualistic thinking
ecofascism
egotism/narrow sense of ego
elitism
emotional fragility/acting spoiled
enmity/antagonism/hostility
environmentalism
envy/covetousness
eros
exclusivism
exploitation
fanaticism
fashionableness
fatalism
fear
foolishness
forcing
forgetfulness/amnesia
fundamentalism/absolutist thinking
gluttony
greed/avarice/acquisitiveness
hardness of heart
hatred
hedonism
heroism
hierarchical sensibility/hierarchical thinking
homophobia
humiliation
hurrying/hastiness
hypocrisy
idealism/utopianism
idolatry
ignorance
immaturity/pseudomaturity
impatience
indifference
individualism/"solo playing"/isolation
insensitivity
instrumentalism
intellectualism/rationalism
intolerance
irrationality
irresponsibility
judgmentalism
lack of discipline
lack of imagination
lack of mercy
lack of respect/devaluation/disregard
lack of rootedness
lack of self-awareness
legalism
lovelessness
lying
manipulation
mastery/domination/oppression
mechanistic thinking
meddling/excessive intervention
militarism
misanthropy
misogyny
murder
naiveté/simple-mindedness
narcissism
nastiness
nationalism
neglect
nobility
obedience
objectification (of animals, nature, women)
opulence
ostentatiousness
otherworldliness
overconfidence/sense of omnipotence
pantheism/monism (spiritual)
parochialism
passivity/lethargy
paternalism
patriarchal thinking
pessimism/gloominess

pragmatism
pride/hubris/self-aggrandizement
prodigality/profligacy
projection
protecting (as hero complex)
punishing
racism
rapaciousness/rape
reductionism
refusal to deal with own waste
repression/suppression
resentment
restlessness
rigidity/inflexibility
romanticism
saving (as hero complex)
scientific attitudes
self-deception/self-delusion/fantasy
self-deprecation
self-doubt
self-indulgence
selfishness/self-centeredness/self-seeking love/egoism
self-righteousness
sentimentality
sexism
shallow thinking/narrow thinking/short-sightedness
slavishness
spiritualism
sloth
speciesism
stupidity/lack of wisdom/foolishness
sullenness
superstition
technocratic mentality/faith in industrial technology
thoughtlessness/mindlessness
triumphalism
trivialization
tyranny
victim-blaming
violence
wastefulness
wantonness
wishful thinking
xenophobia
zeal (in excess)

Appendix B
Nature and Ethics: A Working Typology

\ function image \	IMPETUS	SUPPORT	LIMITATION	DIRECTION	PRESERVATION	CONSTRUCTION	CHOICE
DIVINE POWER	**provides authority**			**reveals vision**		**reveals plan of action**	**reveals choices**
COSMIC EMBRACE				**structures virtue**		**structures virtue**	
BOOK OF LAW			**legislates**	**legislates**			
MATRIX	*provides meaning*	*restores wholeness*		*helps develop ideals*			
PYRAMID							ranks values
COMMUNITY		**nourishes, activates**	**shows limits of survival**	**requires integrated functioning**	**models preserving functions**	**models constructive functions**	
RESPECTED OTHER							ensures care in sacrifice

bold: roles of nature occurring in *The Dream of the Earth*
italics: roles of nature occurring in *The Ecology of Freedom*

BIBLIOGRAPHY

Adams, Carol, ed. *Ecofeminism and the Sacred.* Maryknoll, NY: Orbis Press, 1992.

Aquinas, Thomas. *Summa Theologiae.* Trans. the Fathers of the English Dominican Province. New York: Benziger Brothers, 1948.

Aristotle. *The Nicomachean Ethics.* Trans. W. D. Ross. Oxford, UK: Oxford University Press, 1925; reprint ed., 1990.

Augustine. *Confessions.* New York, NY: Penguin Classics, 1961.

———. *The Catholic and Manichaean Ways of Life.* Trans. D. A. Gallagher and I. J. Gallagher. Washington, DC: Catholic University of America Press, 1966.

Beach, Waldo, and H. Richard Niebuhr, eds. *Christian Ethics: Sources of the Living Tradition.* New York, NY: Ronald Press Company, 1955.

Benton, L. M. "Selling the Natural or Selling Out? Exploring Environmental Merchandising." *Environmental Ethics* 17 (1995): 3–22.

Bergen, Annegreet van, and Marijke Hilhorst. "De Lusthof Nederland." *Elsevier* 25 (22 June 1996): 64–70.

Berry, Thomas. *The Dream of the Earth.* San Francisco, CA: Sierra Club Books, 1988.

Berry, Wendell. *A Continuous Harmony: Essays Cultural and Agricultural.* San Diego: Harcourt Brace Jovanovich, 1972, 1970.

Birch, Charles. "The Scientific-Environmental Crisis: Where Do the Churches Stand?" *The Ecumenical Review* 40 (April 1988): 185–93.

Birch, Charles, William Eakin, and Jay B. McDaniel, eds. *Liberating Life: Contemporary Approaches to Ecological Theology.* Maryknoll, NY: Orbis Press, 1990.

Bloomfield, Morton W. *The Seven Deadly Sins: An Introduction to the History of a Religious Concept, with Special Reference to Medieval English Literature.* East Lansing, MI: Michigan State College Press, 1952.

Bookchin, Murray [Lewis Herber]. "The Problem of Chemicals in Food." *Contemporary Issues* 3, no. 12 (1952): 206–41.

———. "Beyond Neo-Marxism." *Telos* 36 (Summer 1978): 5–28.

———. *Toward an Ecological Society.* Montreal, Canada: Black Rose Books, 1980.

———. *The Ecology of Freedom: The Emergence and Dissolution of Hierarchy.* Palo Alto, CA: Cheshire Books, 1982.

———. *The Modern Crisis.* Philadelphia, PA: New Society Publishers, 1986.

———. *The Rise of Urbanization and the Decline of Citizenship.* San Francisco, CA: Sierra Club Books, 1987.

———. *Remaking Society: Pathways to a Green Future.* Boston, MA: South End Press, 1990.

———. *The Philosophy of Social Ecology: Essays on Dialectical Naturalism.* Montreal, Canada: Black Rose Books, 1990.

Bowman, Douglas C. *Beyond the Modern Mind: The Spiritual and Ethical Challenge of the Environmental Crisis.* New York, NY: Pilgrim Press, 1990.

Brown, Lester R. *Building a Sustainable Society.* New York, NY: W. W. Norton & Company, 1981.

Bruner, Michael, and Max Oelschlaeger. "Rhetoric, Environmentalism, and Environmental Ethics." *Environmental Ethics* 16 (1994), 377–96.

Callicott, J. Baird. *In Defense of the Land Ethic: Essays in Environmental Philosophy.* Albany, NY: SUNY Press, 1989.

———. "The Land Ethic Today." *Topoi* 12 (1993): 41–51.

Callicott, J. Baird, and R. T. Ames, eds. *Nature in Asian Traditions of Thought: Essays in Environmental Philosophy.* Albany, NY: SUNY Press, 1989.

Capps, Donald. *Deadly Sins and Saving Virtues.* Philadelphia, PA: Fortress Press, 1989.

Casey, John. *Pagan Virtue: An Essay in Ethics.* Oxford, UK: Clarendon Press, 1990.

Cessario, Romanus, O.P. *The Moral Virtues and Theological Ethics.* Notre Dame, IN: University of Notre Dame Press, 1991.

Chapple, Christopher Key, ed. *Ecological Prospects: Scientific, Religious, and Aesthetic Perspectives.* Albany, NY: SUNY Press, 1994.

Chase, Steve, ed. *Defending the Earth: A Dialogue Between Murray Bookchin & Dave Foreman.* Boston, MA: South End Press, 1991.

Cheney, Jim. "Ecofeminism and Deep Ecology." *Environmental Ethics* 9 (1987): 115–45.

Christ, Carol P., and Judith Plaskow, eds. *Womanspirit Rising: A Feminist Reader in Religion.* San Francisco, CA: Harper & Row, 1979.

Christiansen, Drew, S. J., and Walter Grazer, eds. *"And God Saw That It Was Good": Catholic Theology and the Environment.* Washington, DC: United States Catholic Conference, 1996.

Cobb, John B., Jr. *Sustainability: Economics, Ecology, and Justice.* Maryknoll, NY: Orbis Books, 1992.

Copeland, Lennie. "Learning to Manage a Multicultural Workforce." *Training* 25 (May 1988): 48–9, 51, 55–6.

Corson, Walter H., ed. *The Global Ecology Handbook.* Boston, MA: Beacon Press, 1990.

Dalton, Anne Marie. "Befriending an Estranged Home." *Religious Education* 85 (Winter 1990): 15–24.

Daly, Mary. *Pure Lust: Elemental Feminist Philosophy.* San Francisco, CA: Harper, 1984.

Devall, Bill. *Simple in Means, Rich in Ends: Practicing Deep Ecology.* Salt Lake City, UT: Gibbs Smith, 1988.

Devall, Bill, and George Sessions. *Deep Ecology.* Salt Lake City, UT: Gibbs Smith, 1985.

Diamond, Irene, and Gloria Orenstein, eds. *Reweaving the World: The Emergence of Ecofeminism.* San Francisco, CA: Sierra Club Books, 1990.

DiZerega, Gus. "Individuality, Human and Natural Communities, and the Foundation of Ethics." *Environmental Ethics* 17 (1995): 23–37.

Ebeling, Gerhard. *Luther: An Introduction to His Thought.* Trans. R. A. Wilson. Philadelphia, PA: Fortress Press, 1970.

Eckersley, Robyn. "Divining Evolution: The Ecological Ethics of Murray Bookchin." *Environmental Ethics* 11 (1989): 99–116.

Ehrenfeld, David. *The Arrogance of Humanism.* New York, NY: Oxford University Press, 1978.

Elshtain, Jean Bethge. *Public Man, Private Woman: Women in Social and Political Thought.* Princeton, NJ: Princeton University Press, 1981.

Erickson, Ron. "Comment on Environmental Virtue Ethics." *Environmental Ethics* 16 (1994): 334–6.

Erikson, Erik H. *Insight and Responsibility: Lectures on the Ethical Implications of Psychoanalytic Insight.* New York, NY: W. W. Norton, 1964.

Fairly, Henry. *The Seven Deadly Sins Today*. Washington, DC: New Republic Books, 1978.
Farley, Edward. *Good and Evil: Interpreting a Human Condition*. Minneapolis, MN: Fortress Press, 1990.
Foot, Philippa. *Virtues and Vices and Other Essays in Moral Philosophy*. Berkeley, CA: University of California Press, 1978.
Fox, Matthew. *Creation Spirituality: Liberating Gifts for the Peoples of the Earth*. San Francisco, CA: Harper, 1991.
Frasz, Geoffrey B. "Environmental Virtue Ethics: A New Direction for Environmental Ethics." *Environmental Ethics* 15 (1993): 259–74.
French, William C. "Ecological Concern and the Anti-Foundationalist Debates: James Gustafson on Biospheric Constraints." *The Annual of the Society of Christian Ethics* (1989): 113–30.
Gaard, Greta, ed. *Ecofeminism: Women, Animals, Nature*. Philadelphia, PA: Temple University Press, 1993.
Gadamer, Hans-Georg. *Truth and Method*. New York, NY: Crossroad, 1975.
Gallagher, John A. *Time Past, Time Future: An Historical Study of Catholic Moral Theology*. New York, NY: Paulist Press, 1990.
Geertz, Clifford. *The Interpretation of Cultures*. New York, NY: Basic Books, 1973.
Gottlieb, Roger, ed. *This Sacred Earth: Religion, Nature, Environment*. New York, NY: Routledge, 1996.
Green, Karen. "Freud, Wollstonecraft, and Ecofeminism: A Defense of Liberal Feminism." *Environmental Ethics* 16 (1994): 117–34.
Griffin, Susan. *Woman and Nature: The Roaring Inside Her*. New York, NY: Harper & Row, 1978.
Grisez, Germain, and Russell Shaw. *Fulfillment in Christ: A Summary of Christian Moral Principles*. Notre Dame, IN: University of Notre Dame Press, 1991.
Gula, Richard. *Reason Informed by Faith: Foundations of Catholic Morality*. New York, NY: Paulist Press, 1989.
Halkes, Catharina. *En Alles Zal Worden Herschapen: Gedachten Over de Heelwording van de Schepping in het Spanningsveld tussen Natuur en Cultuur*. Baarn, The Netherlands: Ten Have, 1989.
Hallman, David G. *A Place in Creation: Ecological Visions in Science, Religion, and Economics*. Toronto, Canada: United Church Publishing House, 1992.
Hare, R. M. *Freedom and Reason*. Oxford: Oxford University Press, 1963.
Häring, Bernard. *The Law of Christ*. Westminster: Newman Press, 1961, 1963, 1966.
———. *Free and Faithful in Christ* vol. 1: *General Moral Theology*. New York, NY: Seabury Press, 1978, 1979, 1981.
Hargrove, Eugene C., ed. *Religion and Environmental Crisis*. Athens, GA: University of Georgia Press, 1986.
———. "The Role of Rules in Ethical Decision Making." *Inquiry* 28 (1985): 3–42.
Hauerwas, Stanley. *A Community of Character: Toward a Constructive Christian Social Ethic*. Notre Dame, IN: University of Notre Dame Press, 1981.
———. *Vision and Virtue*. Notre Dame, IN: Fides Press, 1974.
Hauerwas, Stanley, and Alasdair MacIntyre, eds. *Revisions: Changing Perspectives in Moral Philosophy*. Notre Dame, IN: University of Notre Dame Press, 1983.
Haught, John F. *The Promise of Nature*. New York, NY: Paulist Press, 1993.
Hessel, Dieter, ed. *For Creation's Sake: Preaching, Ecology, and Justice*. Philadelphia, PA: Geneva Press, 1985.
Hill, Thomas E. "Ideals of Human Excellence and Preserving Natural Environments." *Environmental Ethics* 5 (Fall 1983): 211–24.

Himes, Kenneth R. "Social Sin and the Role of the Individual." *The Annual of the Society of Christian Ethics* (1986): 183–218.
Irwin, Terence. *Aristotle's First Principles.* Oxford, UK: Clarendon Press, 1988.
John Paul II. "Sollicitudo Rei Socialis." *Origins* 17 (3 March 1988): 641, 643–60.
———. "The Ecological Crisis: A Common Responsibility." World Day of Peace Message, 1 January 1990. USCC Publication No. 332–9.
Johnson, Lawrence. *Morally Deep World: An Essay on Moral Significance and Environmental Ethics.* Cambridge, UK: Cambridge University Press, 1991.
Jonas, Hans. *The Phenomenon of Life.* New York, NY: Delta Books, 1966.
Joranson, Philip N., and Ken Butigan, eds. *Cry of the Environment: Rebuilding the Christian Creation Tradition.* Santa Fe, NM: Bear & Company, 1984.
Julian of Norwich. *Showings.* Trans. with introduction by Edmund Colledge, O.S.A. and James Walsh, S.J. New York, NY: Paulist Press, 1978.
Kidner, David W. "Why Psychology Is Mute About the Environmental Crisis." *Environmental Ethics* 16 (1994): 359–76.
Levin, D. Michael. "Transpersonal Experience: Practising Deep Ecology in Education and Self-Formation." Paper presented at the 1991 Casassa Conference, Loyola Marymount University, Los Angeles, CA.
Lilburne, Geoffrey R. *A Sense of Place: A Christian Theology of the Land.* Nashville, TN: Abingdon Press, 1989.
Linzey, Andrew. *Christianity and the Rights of Animals.* New York, NY: Crossroad, 1987.
Lonergan, Anne, and Caroline Richards, eds. *Thomas Berry and the New Cosmology.* Mystic, CT: 23rd Publications, 1987.
Luke, Timothy W. "Social Ecology as Critical Political Economy." *The Social Science Journal* 24 (1987): 303–15.
MacIntyre, Alasdair. *After Virtue: A Study in Moral Theory.* 2nd ed. Notre Dame, IN: University of Notre Dame Press, 1984.
Mahoney, John. *The Making of Moral Theology: A Study of the Roman Catholic Tradition.* Oxford, UK: Clarendon Press, 1987.
Marietta, Don E., Jr. "Pluralism in Environmental Ethics." *Topoi* 12 (1993): 69–80.
May, William F. *A Catalogue of Sins: A Contemporary Examination of Christian Conscience.* New York, NY: Holt, Rinehart and Winston, 1967.
McClendon, James, Jr. *Ethics: Systematic Theology* vol. I. Nashville, TN: Abingdon Press, 1986.
McDaniel, Jay B. *Of God and Pelicans: A Theology of Reverence for Life.* Louisville, KY: Westminster/John Knox Press, 1989.
McDonagh, Sean. *To Care for the Earth: A Call to a New Theology.* Santa Fe, NM: Bear & Company, 1986.
McFague, Sallie. *Models of God: Theology for an Ecological, Nuclear Age.* Philadelphia, PA: Fortress Press, 1987.
Meilaender, Gilbert. *The Theory and Practice of Virtue.* Notre Dame, IN: University of Notre Dame Press, 1984.
Menninger, Karl. *Whatever Became of Sin?* New York, NY: Hawthorn Books, 1973.
Merchant, Carolyn. *The Death of Nature: Women, Ecology and the Scientific Revolution.* San Francisco, CA: Harper & Row, 1980.
Mieth, Dietmar, and Jacques Pohier, eds. *Changing Values and Virtues.* Concilium 191 Special Column. Edinburgh, UK: T. & T. Clark, 1987.
Miller, Alice. *The Drama of the Gifted Child.* Trans. R. Ward. New York, NY: Basic Books, 1981.

Moore, G. E. *Principia Ethica*. Cambridge: Cambridge University Press, 1903, 1956.

Naess, Arne. *Ecology, Community and Lifestyle: Outline of an Ecosophy*. Trans. David Rothenberg. Cambridge: Cambridge University Press, 1989.

Nash, James A. *Loving Nature: Ecological Integrity and Christian Responsibility*. Nashville, TN: Abingdon Press, 1991.

———. "Toward the Revival and Reform of the Subversive Virtue: Frugality." *The Annual of the Society of Christian Ethics* (1995): 137–60.

Niebuhr, H. Richard. *The Responsible Self*. New York, NY: Harper & Row, 1963.

Northcott, Michael S. *The Environment and Christian Ethics*. Cambridge, UK: Cambridge University Press, 1996.

Norton, Bryan. *Toward Unity Among Environmentalists*. New York, NY: Oxford University Press, 1991.

———. "Should Environmentalists Be Organicists?" *Topoi* 12 (1993): 21–30.

Oelschlaeger, Max. *Caring for Creation: An Ecumenical Approach to the Environmental Crisis*. New Haven, CT: Yale University Press, 1994.

Okin, Susan Moller. *Women in Western Political Thought*. Princeton, NJ: Princeton University Press, 1979.

Patterson, John. "Maori Environmental Virtues." *Environmental Ethics* 16 (1994): 397–409.

Pieper, Josef. *The Four Cardinal Virtues*. Notre Dame, IN: University of Notre Dame Press, 1966.

Pitkin, Hanna Fenichel. *Fortune Is a Woman: Gender and Politics in the Thought of Niccolò Machiavelli*. Berkeley, CA: University of California Press, 1984.

Plant, Judith, ed. *Healing the Wounds: The Promise of Ecofeminism*. Philadelphia, PA: New Society, 1989.

Plato. The "Republic." In *Five Great Dialogues*. Trans. Benjamin Jowett. Roslyn, NY: Classics Club, Walter J. Black, 1942, 221–53.

Plumwood, Val. *Feminism and the Mastery of Nature*. New York, NY: Routledge, 1993.

Porter, Jean. *The Recovery of Virtue: The Relevance of Aquinas for Christian Ethics*. Louisville, KY: Westminster/John Knox Press, 1990.

Ramsey, Paul. *Basic Christian Ethics*. New York, NY: Charles Scribner's Sons, 1950.

Regan, Tom. *The Case for Animal Rights*. Berkeley and Los Angeles, CA: University of California Press, 1983.

Ricœur, Paul. *The Symbolism of Evil*. Trans. E. Buchanan. New York, NY: Harper & Row, 1967.

Riker, John H. *Human Excellence and an Ecological Conception of the Psyche*. Albany, NY: SUNY Press, 1991.

Roberts, Robert C. "Will Power and the Virtues." *The Philosophical Review* 93 (April 1984): 227–47.

———. "Therapies and the Grammar of Virtue." In *The Grammar of the Heart*, ed. Richard H. Bell. San Francisco, CA: Harper & Row, 1988, 149–70.

Rolston, Holmes, III, "Can and Ought We to Follow Nature?" *Environmental Ethics* 1 (1979): 7–30.

Rorty, Amélie Oksenberg, ed. *Essays on Aristotle's Ethics*. Berkeley, CA: University of California Press, 1980.

Rosaldo, Michelle, and Louise Lamphere, eds. *Woman, Culture, and Society*. Stanford, CA: Stanford University Press, 1974.

Ruether, Rosemary Radford. *New Woman/New Earth: Sexist Ideologies and Human Liberation*. New York, NY: Seabury Press, 1975.

———. "Courage as a Christian Virtue," *Cross Currents* 33 (Spring 1983): 8–16.

———. *Gaia and God: An Ecofeminist Theology of Earth Healing*. San Francisco, CA: HarperCollins, 1992.

Sandilands, Catriona. "From Natural Identity to Radical Democracy." *Environmental Ethics* 17 (1995): 75–92.

Santmire, J. Paul. *The Travail of Nature: The Ambiguous Ecological Promise of Christian Theology*. Philadelphia, PA: Fortress Press, 1985.

Schaefer, Jamie Ehegartner. "Ethical Implications of Applying Aquinas's Notions of the Unity and Diversity of Creation to Human Functioning in Ecosystems." Marquette University: Ph.D. dissertation, 1994.

Schimmel, Solomon. *The Seven Deadly Sins: Jewish, Christian, and Classical Reflections on Human Nature*. New York, NY: Free Press, 1992.

Sherman, Nancy. *The Fabric of Character: Aristotle's Theory of Virtue*. Oxford, UK: Clarendon Press, 1989.

Shiva, Vandana. *Staying Alive: Women, Ecology and Development*. London, England: Zed Books, 1989.

Shklar, Judith N. *Ordinary Vices*. Cambridge, MA: Belknap Press of Harvard University Press, 1984.

Simon, Thomas W. "Varieties of Ecological Dialectics." *Environmental Ethics* 12 (1990): 211–31.

Slack, Kenneth. *The Seven Deadly Sins: A Contemporary View*. London, England: SCM Press, 1985.

Slicer, Deborah. "Is There an Ecofeminism-Deep Ecology 'Debate'?" *Environmental Ethics* 17 (1995): 151–69.

Smith, Pamela A. "Aquinas and Today's Environmental Ethics: An Exploration of How the Vision and the Virtue Ethic of 'Ecothomism' Might Inform a Viable Eco-Ethic." Duquesne University, Ph.D. dissertation, 1995.

Soelle, Dorothee. *The Window of Vulnerability: A Political Spirituality*. Minneapolis, MN: Fortress Press, 1990.

Solomon, Robert. *Ethics and Excellence: Cooperation and Integrity in Business*. New York, NY: Oxford University Press, 1992.

Spretnak, Charlene. *The Spiritual Dimension of Green Politics*. Santa Fe, NM: Bear & Company, 1986.

Stevenson, Brian K. "Contextualism and Norton's Convergence Hypothesis." *Environmental Ethics* 17 (1995): 135–50.

Stone, Christopher. *The Gnat Is Older Than Man: Global Environment and Human Agenda*. Princeton, NJ: Princeton University Press, 1993.

Strehlow, Wighard, and Gottfried Hertzka. *Hildegard of Bingen's Medicine*. Trans. Karin Anderson Strehlow. Santa Fe, NM: Bear & Company, 1988.

Strong, David. "Disclosive Discourse, Ecology, and Technology." *Environmental Ethics* 16 (1994): 89–102.

Suess, Eduard. *The Face of the Earth* vol. 5. Trans. Hertha B. C. Sollas. Oxford, UK: Clarendon Press, 1904, 1924.

Swimme, Brian, and Thomas Berry. *The Universe Story: From the Primordial Flaring Forth to the Ecozoic Era*. San Francisco, CA: Harper, 1992.

Taylor, Paul W. *Respect for Nature: A Theory of Environmental Ethics*. Princeton, NJ: Princeton University Press, 1986.

Tillich, Paul. *The Courage to Be*. New Haven, CT: Yale University Press, 1952.

Trible, Phyllis. *Texts of Terror*. Philadelphia, PA: Fortress Press, 1984.

Tucker, Mary Evelyn. "An Ecological Cosmology: The Confucian Philosophy of Material

Force." Paper presented at the 1991 Casassa Conference, Loyola Marymount University, Los Angeles, CA.

Tussen Heks en Heilige [Between Witch and Saint]: Het Vrouwbeeld op de Drempel van de Moderne Tijd, 15de/16de Eeuw. Nijmegen, the Netherlands: Uitgeverij SUN, 1985.

Uhlein, Gabriele. *Meditations with Hildegard of Bingen*. Santa Fe, NM: Bear & Company, 1982.

Wallace, James D. *Virtues and Vices*. Ithaca, NY: Cornell University Press, 1978.

Weiner, Neil O. *The Harmony of the Soul: Mental Health and Moral Virtue Reconsidered*. Albany, NY: SUNY Press, 1993.

Welch, Sharon D. *A Feminist Ethic of Risk*. Minneapolis, MN: Fortress Press, 1990.

Wensveen, Louke M. van. "Is Toughness a Business Virtue?" *The International Journal of Applied Philosophy* (Winter/Spring 1995): 15–25.

Weston, Anthony. *Back to Earth: Tomorrow's Environmentalism*. Philadelphia, PA: Temple University Press, 1994.

Westra, Laura. *An Environmental Proposal for Ethics: The Principle of Integrity*. Lanham: Rowman & Littlefield, 1994.

White, Lynn, Jr. "The Historical Roots of Our Ecologic Crisis." *Science* 155 (10 March 1967): 1203–7.

Whitehead, James D., and Evelyn Eaton Whitehead. "Christians and Their Passions." *Warren Lecture Series in Catholic Studies* 21. Tulsa, OK: University of Tulsa, 1992.

Whitman, Walt. *Leaves of Grass: The 1892 Edition*. Introduction by Justin Kaplan. New York, NY: Bantam Books, 1983.

Williams, Bernard. *Ethics and the Limits of Philosophy*. Cambridge, MA: Harvard University Press, 1985.

Wilson, Marvin R. *Our Father Abraham: Jewish Roots of the Christian Faith*. Grand Rapids, MI: Eerdmans Press, 1989.

Wogaman, J. Philip. *Christian Ethics: A Historical Introduction*. Louisville, KY: Westminster/John Knox Press, 1993.

Yearley, Lee H. *Mencius and Aquinas: Theories of Virtue and Conceptions of Courage*. Albany, NY: SUNY Press, 1990.

———."Recent Work on Virtue." *Religious Studies Review* 16, no. 1 (January 1990): 1–9.

Zimmerman, Michael E. "Rethinking the Heidegger-Deep Ecology Relationship." *Environmental Ethics* 15 (1993): 195–224.

INDEX